Ahmad Galadima
Zahraddeen Nasiru G. Getso
Mustapha Danlami Garba

Carbon Capture and Storage in the Fight for Sustainable Environment

Ahmad Galadima
Zahraddeen Nasiru G. Getso
Mustapha Danlami Garba

Carbon Capture and Storage in the Fight for Sustainable Environment

Science and Prospects

LAP LAMBERT Academic Publishing

Impressum/Imprint (nur für Deutschland/only for Germany)
Bibliografische Information der Deutschen Nationalbibliothek: Die Deutsche Nationalbibliothek verzeichnet diese Publikation in der Deutschen Nationalbibliografie; detaillierte bibliografische Daten sind im Internet über http://dnb.d-nb.de abrufbar.

Coverbild: www.ingimage.com

Verlag: LAP LAMBERT Academic Publishing GmbH & Co. KG
Heinrich-Böcking-Str. 6-8, 66121 Saarbrücken, Deutschland
Telefon +49 681 3720-310, Telefax +49 681 3720-3109
Email: info@lap-publishing.com

Approved by: Personal Research

Herstellung in Deutschland:
Schaltungsdienst Lange o.H.G., Berlin
Books on Demand GmbH, Norderstedt
Reha GmbH, Saarbrücken
Amazon Distribution GmbH, Leipzig
ISBN: 978-3-8484-1134-4

Imprint (only for USA, GB)
Bibliographic information published by the Deutsche Nationalbibliothek: The Deutsche Nationalbibliothek lists this publication in the Deutsche Nationalbibliografie; detailed bibliographic data are available in the Internet at http://dnb.d-nb.de.

Cover image: www.ingimage.com

Publisher: LAP LAMBERT Academic Publishing GmbH & Co. KG
Heinrich-Böcking-Str. 6-8, 66121 Saarbrücken, Germany
Phone +49 681 3720-310, Fax +49 681 3720-3109
Email: info@lap-publishing.com

Printed in the U.S.A.
Printed in the U.K. by (see last page)
ISBN: 978-3-8484-1134-4

About the Authors

Hailed from Kotorkoshi in Nigeria, Ahmad Galadima received his First Class Bachelor's degree in chemistry from Bayero University, Kano, Nigeria in 2005, and MSc Hydrocarbon Enterprise (with Distinction) from University of Aberdeen, United Kingdom in 2008. He teaches physical and environmental chemistry courses at Usmanu Danfodiyo University, Sokoto, Nigeria. Galadima has won both national and international prizes such as Michael Collin's Prize for Best Graduation Student in Chemistry, 2005 and Bank of Scotland Best Dissertation Prize, 2008. He has good publications record and is currently doing a PhD on upgrading gasoline at the University of Aberdeen.

Zahraddeen Nasiru Garba Getso was born in Getso, Kano, Nigeria. He obtained his BSc degree (second class upper) in chemistry from Bayero University Kano, Nigeria in 2005, and MSc chemistry from Ahmadu Bello University, Zaria, Nigeria in 2010. He lectures environmental chemistry at the latter university and has published over 17 co-authored papers. His PhD research at the same university focused on detection, prevention and control of heavy metals pollution among local communities in Nigeria.

Hailed from Bunkure in Kano state Nigeria, Mustapha Garba Dan-Lami is a dedicated scholar with interest in environmental aspect of oil and gas chemistry. He obtained his BSc chemistry (second class upper) in 2010 from Bayero University Kano, Nigeria. He has contributed to the publication of good number of papers and is currently doing an MSc in oil and gas chemistry at the University of Aberdeen, United Kingdom.

About the Book

Among the climate change mitigation options, carbon capture and storage had been identified to have good future potentials, particularly on the ground that, fossil fuels will remain the major global sources of energy in the medium term. The technology also offers the advantages of enhanced oil and/or gas recovery from partly depleted or low producing reservoirs, and improved methane production from coal deposits. Therefore its full understanding through research and publications as recommended by United Nations Framework Convention on Climate Change (UNFCCC), Intergovernmental Panel on Climate Change (IPCC), Global CCS Institute and many other international organisations is necessary.

The book **"Carbon Capture and Storage in the Fight for Sustainable Environment"** is a relevant material that reports important issues such as history, capture, transportation and storage aspects of the technology. The scope was extended to include commercialisation, ongoing projects and other economics and policy issues. Students and researchers will find it excellent, especially that it’s content, is a summary of wide range and critical survey of relevant literature.

Acknowledgement

All praise is to the CREATOR of here and hereafter. The work reported herein is a combination of the opinions of large number of authors, and associated organisations coupled with the analytical details by the book authors. Their immense contributions to the understanding and dissemination of CCS knowledge, in line with global fight for sustainable environment MUST be acknowledged.

Both the authors and the publisher wished to extend their profound gratitude to Petroleum Technology Development Fund (PTDF), Nigeria for PhD research funding to Ahmad Galadima.

Table of Contents

CHAPTER ONE: INTRODUCTION AND HISTORICAL PERSPECTIVE

1.1 Introduction

For many decades now, fossil fuels (coal, petroleum and natural gas) have accounted for the major source of energy across the globe. There are currently 847 bt, 1,342.207 bbls and 6,254.364 tcf of respective global coal, oil and natural gas reserves (figures 1 to 3), indicating that the fuels will continue to provide energy supply for many more decades to come. However, the numerous challenges associated with their explorations, production and subsequent utilisation necessitate the search for most appropriate ways of sustainable exploitation or alternative energy sources. They are generally non-renewable, and non-evenly distributed, with some countries having very large reserves while others having none.

A key issue that remain the major subject of concern is their associated environmental threats. About 1.30 million tonnes of oil is spilled into the sea annually, killing marine plants and animals and reducing the possibility of fishing in addition to the financial inconsistencies. There are currently over 750 Giga tonnes of carbon dioxide in the atmosphere, resulting from the combustion of these fossil fuels. The carbon dioxide emissions have triggered severe environmental and biological challenges requiring immediate and sustainable actions. The entire globe is alarmingly warming, plant and animal species are disappearing in many regions, level of sea ice is decreasing, and the ocean is becoming acidic. Food production is declining and hunger threats have been observed, especially in the developing countries, where most people leave below an American dollar daily.

To address the problem, several options have been identified, by researchers, companies and international environmental, health and food organisations.

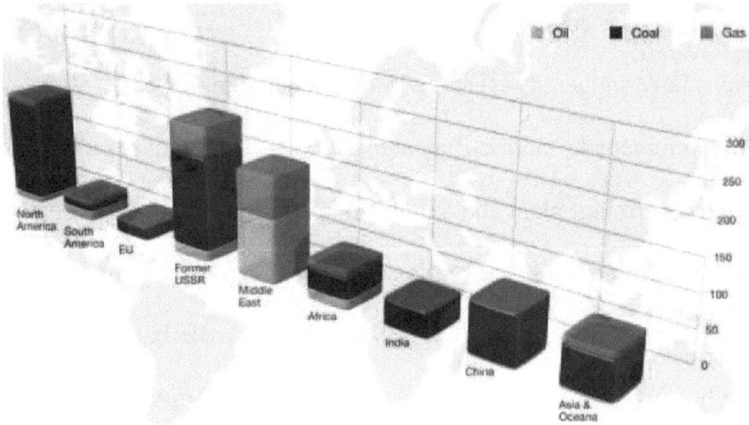

Figure 1: Global coal reserves according to regions (billion tonnes).

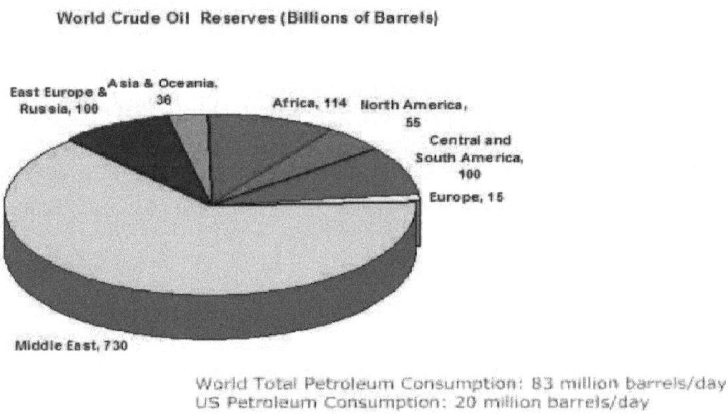

Figure 2: Estimated global proven oil reserves (billion barrels). Errors due to estimation were neglected.

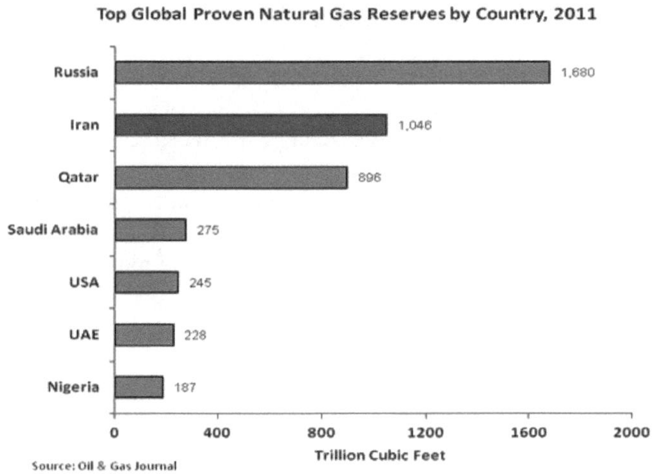

Figure 3: Estimated global proven natural gas reserves by oil and gas journal. Data from non-top countries had been excluded, but could be extracted from the values given in the text.

1.2 Low Carbon Options for Mitigating Climate Change

The key alternatives available for mitigating the increasing challenges of climate change and global warming associated with fossil fuels utilisation are the "low carbon options", which include renewable energy, nuclear power and the carbon capture and storage (CCS).

The growing needs to reduce carbon emissions are driving increasing growth in the **renewable energy** industries. About 85 countries now have targets for their own renewable energy futures, and have enacted wide-ranging international policies to promote renewables production and utilization. Low-carbon renewable energy replaces conventional fossil fuels in three main areas: power generation, hot water/ space heating, and transport fuels. Researchers have prompted a plan to power 100% of the global energy with wind, hydroelectric, and solar power by the year 2050.

In terms of power generation, renewable energy currently provides 18 percent of total electricity generation worldwide and this percentage is growing annually. Renewable power generators are spread across many countries, and wind power alone already provides a significant share of electricity in some areas: for example, 14 percent in the U.S. state of Iowa, 40 percent in the northern German state of Schleswig-Holstein, and 20 percent in Denmark. Some countries get most of their power from renewables, including Iceland (100 percent), Brazil (85 percent), Austria (62 percent), New Zealand (65 percent), and Sweden (54 percent).

Solar water heating makes an important and growing contribution in many countries, most notably in China, which now has 70 percent of the global total (180 GWth). Worldwide, total installed solar water heating systems meet a portion of the water heating needs of over 70 million households. The use of biomass for heating continues to grow as well. In Sweden, national use of biomass energy has surpassed that of oil. Direct geothermal heating is also growing rapidly.

Renewable biofuels for transportation, such as ethanol fuel and biodiesel, have contributed to a significant decline in oil consumption in the United States since the last decade. The 93 billion liters of biofuels produced worldwide in 2009 replaced the equivalent of an estimated 68 billion liters of gasoline, equal to about 6 percent of world gasoline production.

Biofuel use will continue to represent an important share of global cereal, sugar and vegetable oil production over the next decade. By the year 2020, 12% of the global production of coarse grains will be used to produce ethanol compared to 11% on average over the 2008-10 periods. 16% of the global production of vegetable oil will be used to produce biodiesel compared to 11% on average over the 2008-10 period and 33% of the global production of sugar compared to 21% on average over the 2008-10 period (figures 4 and 5). Over the projection period, 21% of the global coarse grains production's increase, 29% of the global vegetable oil production's increase and 68% of the global sugar cane production's increase are expected to go to biofuels (see also figures 4 and 5).

Biogas production is also increasingly gaining attention, with billion of households utilizing the technology worldwide. In the rural China alone there are about 50 million households using the biogas option as a method of energy supply for cooking, heating and lighting. The raw-materials required for the biogas production are abundant. According to the study by Chinese Academy of Sciences, Geography and Resources, China Natural Resources and Environment Data Bank, the total annual production of manure and night soil could theoretically generate about 130 billion m^3 of biogas, equivalent to 93 million tonnes of coal. While only 50 percent of the theoretical production can be realised in rural areas, 80 percent of the industrial wastewater can also be used to produce biogas. By the year 2006 about 1213 GWh (energy equivalent) of biogas production was achieved in Sweden, with the figure increasing rapidly over the years. In the European continent in general, biogas is extensively used for boilers, internal combustion engines, gas turbines, CHP applications, fuels cells and industrial applications. Numerous sources of feedstock have been identified, and with energy policies being modified to meet the full biogas utilisation target. Table 1 reports an estimated energy potential of only pig and cattle manure for the E-27 countries. The data indicate that, biogas has a good future energy prospect in the region. Similar trend has been observed for regions like Argentina, Canada, Brazil and the United States of America. The developing African countries, India and other countries in the Asia have also intensified efforts toward biogas and other biofuels production, as replacement for fossil fuels.

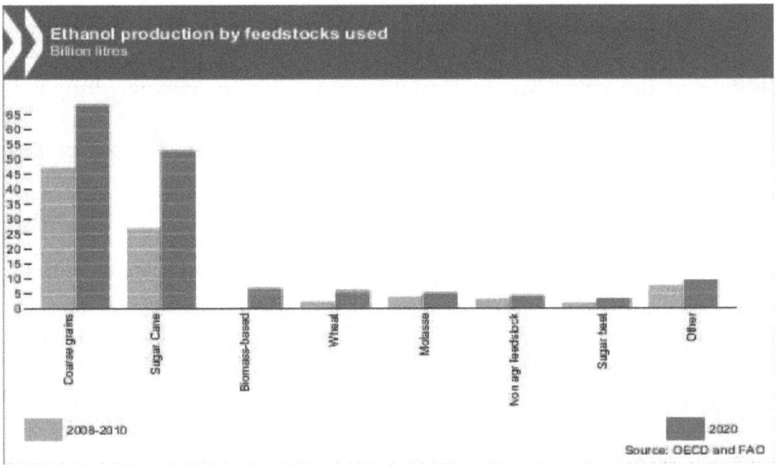

Figure 5: Current and projected global ethanol production according to feedstocks. Source: http://www.agrioutlook.org/document/0/0,3746,en_36774715_36775671_47877696_1_1_1_1,00.html .

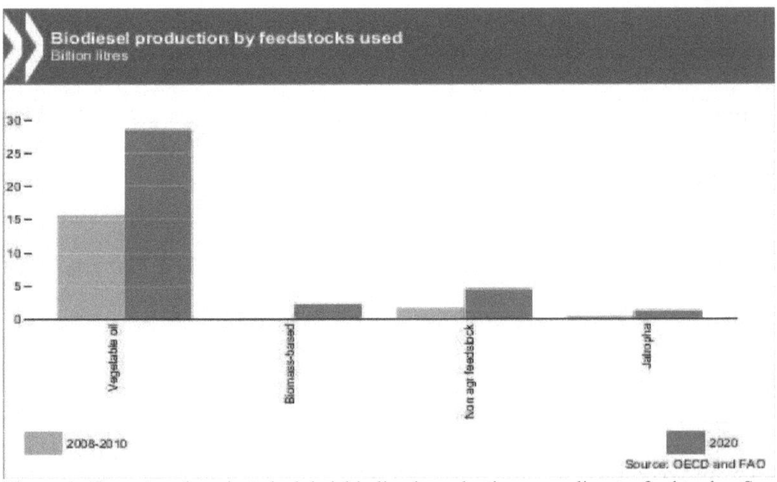

Figure 4: Current and projected global biodiesel production according to feedstocks. Source: http://www.agrioutlook.org/document/0/0,3746,en_36774715_36775671_47877696_1_1_1_1,00.html .

Total Manure [10^6 tons]	Biogas [10^6 m^3]	Methane [10^6 m^3]	Potential [PJ]	Potential [Mtoe]
1,578	31,568	20,519	827	18.5

Table 1: Biogas potential of pig and cattle manure for the E-27 countries, Methane heat of combustion: 40.3 MJ/m^3; 1 Mtoe = 44.8 PJ. Assumed methane content in biogas: 65%. Source: http://www.ramiran.net/doc07/Biogas%20III/Jens_Bo_Holm-Nielsen.pdf

Although the various alternative energy sources have potentials for mitigating energy and environmental inconsistencies associated with the fossil fuels, there are key challenges that may significantly limit their exploitation. Solar power is actually one of the cleanest methods of energy production known. Because solar panels simply convert the energy of the sun into energy mankind can use, there are no harmful byproducts or threats to the environment. One major concern is the cost of solar power. Solar panels (accumulators) are not cheap; and because they are constructed from fragile materials (semiconductors, glass, etc.), they must constantly be maintained and often replaced. Further, since each photovoltaic panel has only about 40% efficiency, single solar panels are not sufficient power producers. However, this problem has been offset by the gathering together of many large panels acting in accord to produce energy. Although this setup takes up much more space, it does generate much more power.

On the other hand, a key disadvantage of using biofuels produced from agricultural crops involves additional land use, as land area is taken up and various agricultural inputs with their environmental effects are inevitable. Switching to biofuels on a large scale requires considerable use of our arable area. Even modest usages of biofuels would consume almost all cropland in some countries. If the same thing is to happen all over the world, the impact on global food supply could be a major concern, and could make some countries being net importers of food products, from their current status of net exporters. It could so happen that most lands on the planet are deployed to produce food for cars, not people (http://www.bdpedia.com/biodiesel/alt/alt.html). Studies have indicated that, already more than 1 billion people are sleeping with hunger across the globe daily. This great challenge will certainly eclipse when food-to-fuel production is intensified.

Another major issue with biodiesel (among other biofuels) is that it tends to reduce fuel economy. Energy efficiency is the percentage of the fuel's thermal energy that is delivered as engine output, and biodiesel has shown no significant effect on the energy efficiency of any test engine. The energy content per gallon of biodiesel is approximately 11 percent lower than that of petroleum diesel. Vehicles running on biodiesel are therefore expected to achieve about 10% fewer miles per gallon of fuel than petrodiesel.

The second option giving consideration is the **nuclear power,** which currently produces 13-14% of the global electricity supply. For over a decade now, the term nuclear renaissance has been used to refer to a possible nuclear power industry revival, driven by rising fossil fuel prices and new concerns about meeting greenhouse gas emission limits. At the same time, various barriers to a nuclear renaissance have been identified. These barriers include unfavorable economics compared to other sources of energy and slowness in addressing climate change.

The new nuclear reactors under construction in Europe, specifically Finland and France, which were meant to lead a nuclear renaissance, have been delayed and are running over-budget. China has twenty new reactors under construction, and there are also a considerable number of new reactors being built in South Korea, India, and Russia. At least 100 older and smaller reactors will "most probably be closed over the next 10-15 years".

Nuclear power has good potentials for waste disposal, safety, and security risks which are unique among low-carbon energy sources. However, public concerns towards nuclear power remain ambiguous in many developed countries, with significant anti-nuclear opposition even when majority opinion is in favour. In the developing African countries, technology remain the major challenge with nuclear energy, and therefore the option has both previously and in the recent years projected to have limited success in the region. The policies designed by the African leaders are similarly not favouring nuclear technology either due to lack of coordinated commitment or corruption and funding problems.

Now, the numerous negative issues attributed to the alternative energy sources coupled with the increased projected energy demand (figure 6) indicated that, the fossil fuels are certainly unavoidable for the medium term. The major issue therefore remains how these fuels could be utilised in the future without compromising environmental sustainability. Since the last two decades, researchers in the relevant fields and industry participants have identified **Carbon Capture and Storage (CCS)** to be a potential alternative. The process involved the isolation, transportation and permanent storage of the CO_2 gas emanating from fossil fuels combustions, without allowing it to reach the atmosphere or seas. This implies that, the fossil fuels would continue to be utilised as energy sources, but the associated CO_2 being removed before reaching the environment. Subsequent sub-sections and chapters of this book will report important details of the CCS technology.

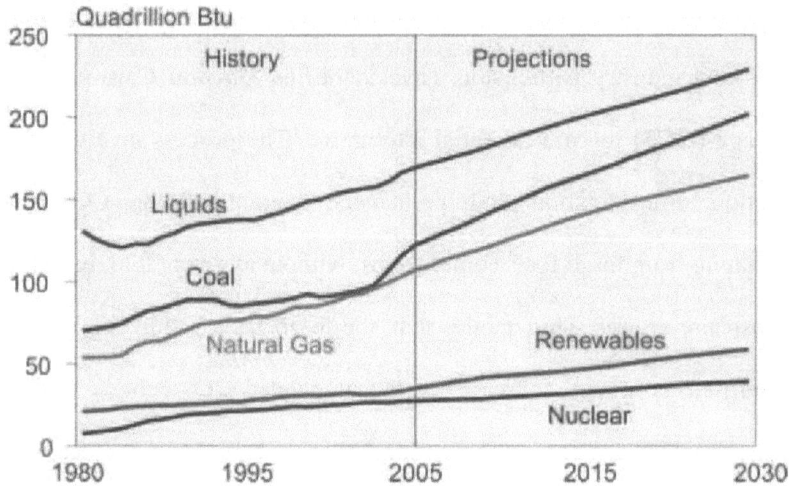

Figure 6: Previous, current and project global energy demand (1980-2030). Source: Energy Information Administration (IEA data), (2007).

1.3 History of Carbon Capture and Storage

The history of CCS as a climate change mitigation option could be traced back to 1970's when pipelines became fully operational for the transportation of captured CO_2 in the Permian Basin of Texas, as an option for Enhanced Oil Recovery (EOR). The technology was by then considered only as a way of improving company's cash flow, through injecting CO_2 and recovering more oil or gas from low producing or partly depleted reservoirs. However, the formal CCS declaration as sustainable environmental technology is in most literatures attributed to the response by the United Nation Framework Convention on Climate Change (UNFCCC), dated forward to 2001, when the seventh conference of parties took place.

In an immediate and active response to the call by the parties, the Intergovernmental Panel on Climate Change (IPCC) planned in 2003 to start a special work, involving the development of comprehensive and special report on the CCS technology. In September 2005, the IPCC working group III approved the report, which also contained useful summary for policy and decision makers, it was later released for public access in the same month.

In the late 2005 the Kyoto Protocol meeting, which was the first Meeting of the Parties, concluded with a decision requesting the UNFCCC Secretariat to arrange a workshop to consider CCS as a clean development technology under the "clean development mechanism (CDM)". The meeting also passed that; the CDM Board Members should devise proposals for new methodologies for CSS and to make key important recommendations to the next Parties Meeting of

November 2006. During the latter convention, the Parties requested the UNFCCC's Subsidiary Body for Scientific and Technological Advice (SBSTA) to prepare decisions and recommendations on CCS in geological formations as CDM project activities for deliberations during future meetings. In the subsequent years, several issues have been considered by all the parties involved. These include, but are not limited to, the followings.

- Long-term physical leakage (seepage) levels of risks and uncertainty; project boundary issues (such as reservoirs in international waters, several projects using one reservoir) and projects involving more than one country (projects that cross national boundaries).
- Long-term responsibility for monitoring the reservoir and any remediation measures that may be necessary after the end of the crediting period.
- Long-term liability for storage sites.
- Accounting options for any long-term seepage from reservoirs.

- Criteria and steps for the selection of suitable storage sites with respect to the potential for release of greenhouse gases.
- Potential leakage paths and site characteristics and monitoring related infrastructure; operation of reservoirs (for example, well-sealing and abandonment procedures), dynamics of CO_2 distribution within the reservoir and remediation issues.
- Any other relevant matters, including environmental impacts.

During the meetings of December 2007, December 2008 and June 2009 several debates took place on the final inclusion of CCS in geological formations as CDM project. Currently, the technology has received widespread in the world and is practised in many facilities.

A survey by the Global CCS Institute reported 74 large-scale integrated CCS projects around the world, 14 of which are either operating or under construction. The total CO_2 storage capacity of these 14 projects is over 33 million tonnes a year. This is broadly equivalent to preventing the emissions from more than six million cars from entering the atmosphere each year. Research is ongoing in different parts of the world, involving links between academic institutions, oil industry and international organizations. However, the CCS technology is still an unknown issue to many people, especially in the developing world. This should not be the case, because the climate change consequences are certainly being felt by all (including non- CO_2 emitters), with the severe impacts dominant in these developing countries.

1.4 Book Objectives

Energy insecurity, climate change and mitigation options are critical issues that should be understood by every person living on earth. Their negative effects are severe and fast growing at an alarming rate, suggesting that, appropriate decisions and subsequent actions are necessary. Our activities **MUST** be shifted toward ensuring an environment that meets our own need without compromising the ability of future generations to meet their own environmental needs (i.e sustainable environmental development). In line with these facts, continuous energy related environmental studies and research, including situation awareness, are unavoidable. Books reporting up to date issues

MUST be available to students, industries, researchers and even the common man in simple and affordable form. Therefore, the book **"CARBON CAPTURE AND STORAGE IN THE FIGHT FOR SUSTAINABLE ENVIRONMENT"** remain relevant.

1.5 References and Further Readings

AG, (2008).Talk About Carbon Capture and Storage. Alberta Government, July2008. http://www.energy.gov.ab.ca/Org/pdfs/FactSheet_CCS.pdf

Alberts, W.W., (1989).The experience curved doctrine reconsidered. Journal of Marketing 53.

Apt, J.,etal., (2007) . Incentives for near-term carbon dioxide geological sequestration. Carnegie Mellon Electricity Industry Centre, October 2007.

Arrow, K.J.,(962). The economic implications of learning by doing. The Review of Economic Studies 29(3). Ball, A.,etal., 2004.The Asia Pacific LNG market :issues and outlook. ABARE Research Report 04.1,Canberra, November2004.

Berkovitz, D.M., (1989) .Price–Anderson Act: model compensation legislation?—The sixty-three million dollar question. Harvard Environmental Law Review 13.

Boucher, R., (2008).Carbon capture and storage early deployment act(A Bill).Office of CongressmanRickBoucher,June2008. BP, 2008. BP statistical review, 2008.

Bowen, F (2011) Carbon capture and storage as a corporate technology strategy challenge. Energy Policy 39 (2011) 2256–2264

CCSB (2009) Carbon capture and storage bulletin. A summary of the High-level conference on fighting climate change with carbon capture and storage (CCS). Published by International Institute for Sustainable Development (IISD), Volume 163, No. 1, June 2009.

Galadima, A. and Garba,, Z.N (2008) Carbon capture and storage in Nigeria: Fundamental science and potential implementation risks. Science World Journal, Volume 3, Number 2, 95-99.

Global CCS Institute (2012) What is CCS?: http://www.globalccsinstitute.com/ccs/what-is-ccs .

IPCC (2012) Intergovernmental Panel on Climate Change, Working Group III. Climate Change Mitigation. http://www.ipcc-wg3.de/publications/special-reports/special-report-on-carbon-dioxide-capture-and-storage/?searchterm=carbon capture and storage

Oltra C., R. Sala , R. Sola`, M. Di Masso , G. Rowe (2010) Lay perceptions of carbon capture and storage technology. International Journal of Greenhouse Gas Control 4 (2010) 698–706.

Terwel, B., Harinck, F., Ellemers, N., Daamen, D., (2006). Just say what they expect you to say: the influence of argumentation on trust in organizations. In: 8th International Conference on Greenhouse Gas, Noruega.

Tokushige, K., Akimoto, K., Tomoda, T. (2007). Public perceptions on the acceptance of geological storage of carbon dioxide and information influencing the acceptance. International Journal of Greenhouse Gas Control 1, 101–112.

Total, (2008). Lacq Basin CO_2 capture and storage pilot project: results of stakeholder consultation. http://www.total.com.

UNFCCC (2008). On accounting of emissions and assigned amount. Kyoto Protocol Reference Manual. United Nation Framework

Convention on Climate Change, 2008.

http://unfccc.int/resource/docs/publications/08_unfccc_kp_ref_manual.pdf

Varun R., DavidG.V., MarkC.T (2010) Carbon capture and storage at scale: Lessons from the growth of analogous energy technologies. Energy Policy 38 (2010) 4089–4098.

VanNoorden, R., (2010). Carbonsequestration:buried trouble. Nature 463,871–873.

Vergragt, P.J. (2009). CCS in the Netherlands: glass half empty or half full? In: Meadow croft, Langhelle (Eds.), Caching the Carbon: The Politics and Policy of Carbon Capture and Storage .Edward Elgar, Cheltenham,UK, pp. 186–210.

World CoalInstitute, (2009). CCS and the CDM. Climate Policy Paper 3. London,UK.

World Energy Council, (2007).Carbon Capture and Storage: AWEC''Interim Balance''. London: UK.

CHAPTER TWO: CAPTURE TECHNOLOGIES

2.0 Carbon (IV) Oxide Capture

Daily concentration of carbon (iv) oxide in the atmosphere is increasing and promoted due to activities of humans, such as combustion of fossils fuel for generation of electricity and also been dependant source of energy. About 80% of the global primary energy comprise of carbon-base fossil fuel resources (Statistical Review, 2004). The CO_2 gas is recognised as the most important anthropogenic greenhouse gas with a current atmospheric concentration of about 750 gigatones, with studies revealing this to remain in the atmosphere for hundreds of years (Lewis, 2007; GF, 2008).

Primarily, the fossil fuel resources are converted to carbon (IV) oxide and release to the atmosphere due to electricity generation, transportation and industrial activities. These would keep on accumulating in the atmosphere substantially to a high level (Keeling and Whorf, 2005) and as a result the level of sea ice and snow cover seriously decreased by about 15% and the world is becoming warmer, with increased spread of diseases and emergence of drought leading to the desertification of fertile soil (Rosenzweig and Casassa, 2007), unless a significant changes are made (figures 7 and 8).

Carbon capture and storage (CCS) is a process by which the emitted carbon (iv) oxide, mainly form hydrocarbon combustion, such as burning of coal, oil and gas is removed before release and pollute the natural atmosphere, and then appropriately stored by the use of this technology. The proven CCS technologies are applied and practice in so many places, such as USA, Europe and Asia (IEA 2006, Iron et al 2007), but yet, the understanding of the technologies is very poor in many countries, especially Africa, despite the contribution of the African countries to global CO_2 emissions (Galadima and Garba, 2008), with Nigeria in the continent being one of the world biggest gas flarers and largest source of carbon (iv) oxide emission from Sub-Saharan Africa (EIA, 2006; Malumfashi, 2007).

2.1 Carbon (IV) Oxide Capture Techniques

Sources producing CO_2, in large amount would allow economies of scale in carbon separation and simplify transportation to a storage site.

Thus, electricity generation and some large fossil fuel burning industrial sources are therefore good candidates for carbon capture. A coal-fired power plant produces a flue gas stream of about 9-14% mole fraction CO_2, while flue gases produce from natural gas with combined-cycle plant typically have a 4% CO_2 content (Davison et al, 2001). It is also advisable for industries producing in large-scale combustion similar to that used for electricity to adopt similar carbon capture techniques. Up to about two-thirds of the CO_2 produce in oil and gas refineries activities are from fossil fuel fired heaters, also about 60% of the CO_2 produce by the iron and steel industry is from blast furnaces (Davison et al, 2001). Flue gases produced as a result of cement production contain about 14-33% mole fraction CO_2 also due to the combustion of fossil fuels for heat and production from calcination (Lyngfelt et al, 2004).

Carbon capture is therefore a process or any technique which removes or reverses the release of CO_2 to the atmosphere, emanating from large scale emissions (i.e stationary or non-mobile sources), and as a result converts the carbon into a viable carbon sink. This is then transported and stored in secure geological formations deep underground. These geological formations include depleted oil and gas field, deep saline aquifers, Ocean Sequestration, deep un-mineable Coal Seams, and natural underground reservoirs.

Proposed technologies or methods involved capturing CO_2 directly from the air to offset emissions elsewhere, or putting carbon containing compounds into a material sink. The majority of carbon capture work focuses on separating CO_2 from the exhaust gas of fossil fuel conversion facilities for potential geologic storage. There are three main techniques for capturing CO_2 from their point sources, namely; Post-combustion, Pre-combustion and Oxyfuel-combustion.

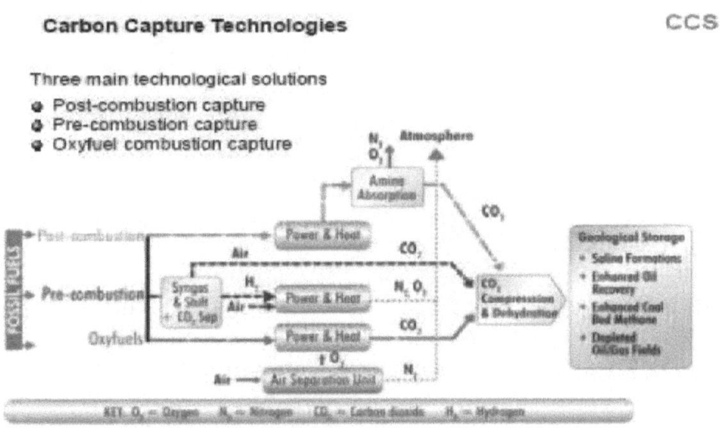

Figure 7: Illustration for the CCS technologies.

The choice of the capture technology or plan is dependent on the region, policies, and level of technology versus emissions. For example, in April 2009, the UK government announced (Post note, 2009) that all new combustion power stations producing 300 MWe or above will have to be 'carbon capture ready' (CCR). This means UK proposals will have to:

• leave sufficient space to retrofit CCS technology, including the necessary pipework

• identify a practical capture option and offshore storage area

• plan a plausible transport route for the CO_2 from the power plant to the storage site

• carry out a technical assessment to ensure there are no foreseeable barriers to retrofitting CCS.

In addition, applicants are also required, under the EU Directive on the Geological Storage of CO_2, to carry out economic assessments on the feasibility of retrofitting CCS and on the transport of CO_2 to the storage area. However, due to economic and technical uncertainties surrounding

CCS, 'capture readiness' cannot guarantee that retrofitting will take place, but does try to ensure that foreseeable barriers are removed.

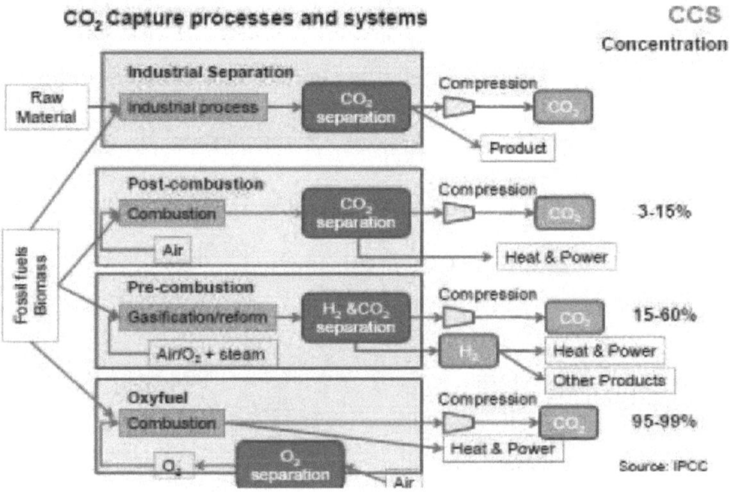

Figure 8: Illustration involving the capture process and systems for the techniques.

2.2 Post-combustion Capture

Post-combustion method involves the removal of the dilute CO_2 from flue gases after hydrocarbon combustion, just before its release to the atmosphere. The method can be typically built into an existing industrial plants and power stations in a process known as retro-fitting, without significant modifications had been made to the original plant. This type of technology is favoured by the UK Government in its competition for state support.

The method uses a physical, chemical or intermediate solvent which can effectively interact with CO_2, stable under desired conditions of temperature and pressure and can easily be regenerated by the shift of these parameters (Galadima and Garba, 2008). There are several ways that can be used to capture the CO_2 in this method.

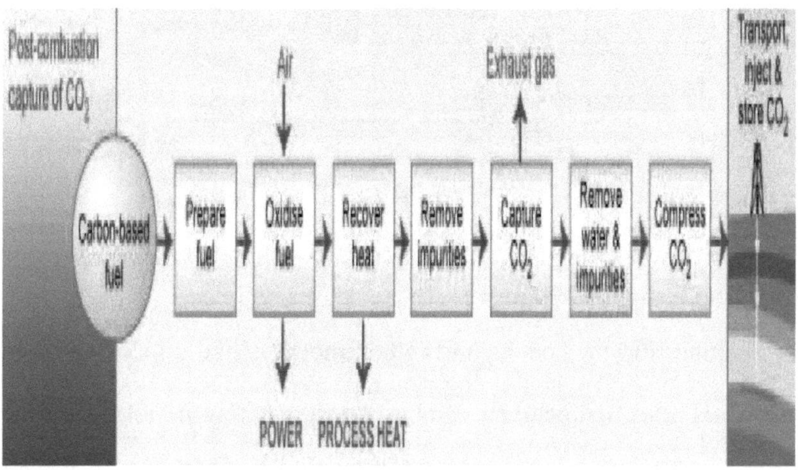

Figure 9: process involving post-combustion capture.

2.2.1 Post Combustion Options

1. Use of Amine

This is the most widely used post combustion option, which involves passing the CO_2 through an amine solvent and adsorbed it. The amine chemical solvents usually in alkanolamines forms used industrially for natural gas sweetening are considered the most likely option for the current world (Galadima and Garba, 2008). The solvents show limited dependency on partial pressure of CO_2, which is normally 15% by volume. The solvent form quaternary Nitrogen carbonate with CO_2, which on heating in the reaction chamber dissociates to release the gas for transportation and the solvents regenerated (Galadima and Garba, 2008). The capacity of the amine to uptake CO_2 is of fundamental importance for sizing the separation tower.

- Aliphatic amine:- are known to react promptly with CO_2 to afford an ammonium carbamate

$$2\ RNH_2 + CO_2 \quad \rightarrow \quad RNH_3^{+-}O_2CNHR$$

At the end of the reaction a cloudy appearance due to a fine suspension of a white solid is obtained (Angela et al 2002). Interestingly, if the reaction is carried out at 273 K, the amount of CO_2 fixed by the amines, after a fast initial uptake, of ca. 0.50 mol/mol, increases slowly to 1 mol/mol within three hours. This was explained by Aresta et al, (2000) Aresta et al (1985) assuming that at low temperature any of the ammonium carbamates RNHCOO-+H3NR slowly converts into the dimeric form of the relevant carbamic $(RNHCOOH)_2$ acid.

$2RNH_2 + CO_2 + H2O \rightarrow (RNH_3)_2CO_3$ (primary amine)

$(RNH_3)_2CO_3 + CO_2 + H_2O \rightarrow 2RNH_3HCO_3$ (secondary amines)

Secondary amines take up CO_2 with a 0.8-1 molar ratio, to afford a glassy material that occludes solvent (Angela et al 2002). The resulting carbamates have been isolated and characterised and thus, far more efficient than primary amines for CO_2-capturing. This is quite important as by using secondary amines, it would be possible to reduce to half the amount of amine necessary for the separation of a given amount of CO_2, and thus the volume of the absorption tower.

Name of chemical	Abbreviation	Scientific formula
Monoethanolamine	MEA	$HOC_2H_4NH_2$
Methyldiethanolamine	MDEA	$(HOC_2H_4)_2NCH_3$
Diethanolamine	DEA	$(HOC_2H_4)_2NH$
Triethanolamine	TEA	$(HOC_2H_4)_3N$
Di-isopropanolamine	DIPA	$(i\text{-}HOC_2H_6)_2NH$

Table 2: list of some common alkanolamines for potential used in CCS. (Galadima and Garba, 2008).

- Aromatic amine

Aromatic amine can also be used in the same conditions, from the carbamate/carbamic acid conversion with benzylamine and other N aromatic compounds.

$$RNHCOO^{-+}H_3NR + CO_2 \rightarrow (RNHCOOH)_2 \qquad R = \text{Benzyl}$$

Carbamates of primary and aromatic amines do not release CO_2 easily upon heating due to its stability and effectiveness only at low concentrations (Galadima and Garba, 2008). Therefore, they are not good candidates for the separation of carbon dioxide from other gases. However, the carbamates of secondary amines easily release carbon dioxide at moderate temperatures, making such systems quite interesting candidates for CO_2 separation.

A change in temperature and or pressure will then release the CO_2 leading to regeneration of the solvent and carbon dioxide release for transportation to appropriate geological storage (figure 10). Alternatively, the process can obey the following mechanism.

$$2RNH_2 + CO_2 \rightarrow RNHCOONH_3R \rightarrow 2RNH_2 + CO_2$$

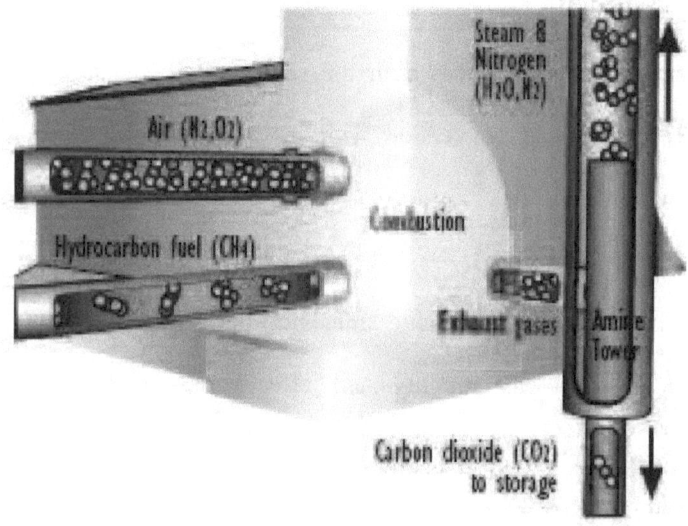

Figure 10: Shows how exhaust gas pass through amine tower where CO_2 is dissolved (Scottish, 2012).

Immediately after the combustion the flue gas is pass through the amine solvent and then dissolve in it, providing a carbon dissolve amine compound known as carbamates, and then to the storage as shown in the figure 10 above.

2. Use of quicklime.

This is another process in development, known as calcium cycle capture. The process involves the use of quicklime to capture the CO_2 and produce a limestone,

$$CaO + CO_2 \rightarrow CaCO_3$$

This can then be heated to drive off the CO2 and recycled back the quicklime which can then be re-used in another subsequent process.

$$CaCO_3 \rightarrow CaO + CO_2$$

All of these require additional energy input to drive off the CO_2 from the solvent and typically results in extra energy costs of 20-30% compared to plants with no capture (Scottish, 2012). New solvents are under development to reduce these penalties to 10%.

Other post-combustion possibilities, currently being researched, include cryogenically solidifying the CO_2 from the flue gases, or removing CO_2 with an adsorbent solid, or by passing CO_2 through a membrane.

Advantages of Post combustion Capture

- It is feasible to retrofit into a current industrial plants and power stations.
- It is an existing technology, over 60 years of experience with amine solvents, but needs 10x scale-up.
- It is currently in use to capture CO_2 for soft drinks industry.

Disadvantages of Post combustion Capture

- High running costs, absorber and degraded solvents replacement.
- Limited large scale operating experience.

2.3 Pre- combustion Capture

Pre- combustion method involves the removal of CO_2 prior to combustion, to produce hydrogen. The hydrogen combustion does not emit CO_2, but with water vapour as the main by-product. Provided that the technology is couple with carbon storage it could provide a CO_2 emission free fuel for the future.

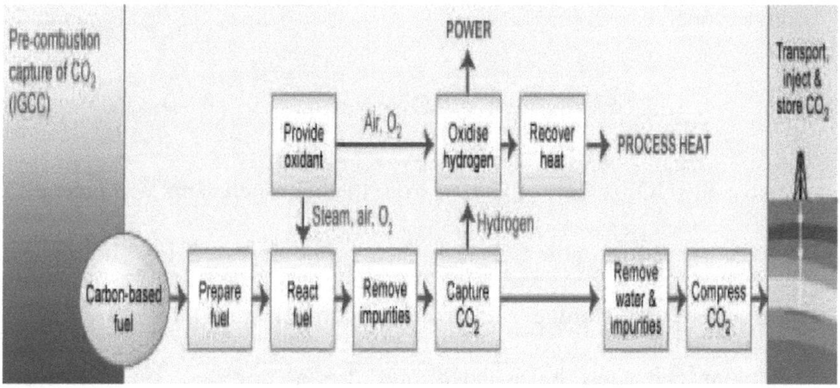

Figure 11: Process involving Pre-combustion (Scottish, 2012).

Pre-combustion method involves three key stages;

Firstly, it involves the conversion of the hydrocarbon fuel (e.g methane, or gasified coal) into hydrogen and carbon monoxide (CO) to form a synthesis gas. This involves the reactions of the fuel in air or oxygen.

$$CxHy + x/2O_2 \rightarrow y/2\ H_2 + xCO \quad \text{(partial oxidation)}$$

$$CxHy + xH_2O \rightarrow xCO + (x+y/2)H_2 \quad \text{(steam reforming)}$$

Secondly, it involve the conversion of the CO into CO2, this is achieve by reacting the gas with steam. The process is known as shift conversion.

$$CO + H_2O \quad \rightarrow \quad CO_2 + H_2$$

Finally, the CO_2 is then separated from the hydrogen. This will provide a clean combustion. The CO_2 can then be compressed into liquid and transported to a storage site and the hydrogen gas, after an effective treatment can then be isolated and use as turbine combine cycle (Galadima and Garba 2008).

Process:

Natural gas, liquid gas and naphtha are generally converted by catalytic steam, depending on the desired final product. In obtaining pure hydrogen as the final product, the process should include a feature such as the catalytic conversion of the CO and a pressure swing adsorption unit (figure 12). This removes all impurities in a single step.

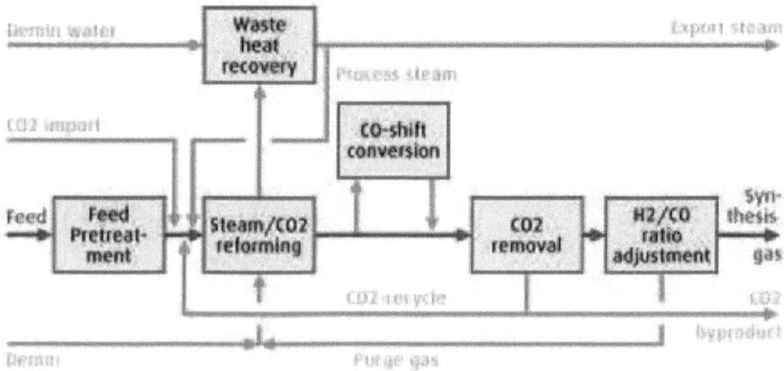

Figure 12: A process involving catalytic converter (Linde 2012).

Heavy oil such as asphalt as well as coal, in this process the oil is partially combusted with oxygen; this is achieved in a non-catalytic partial oxidation producing a raw gas, which is then further processed into saleable products. The process involves a step CO conversion, and other gases are removed with a wash and membrane separation of H_2 and CO is achieved (figure 13).

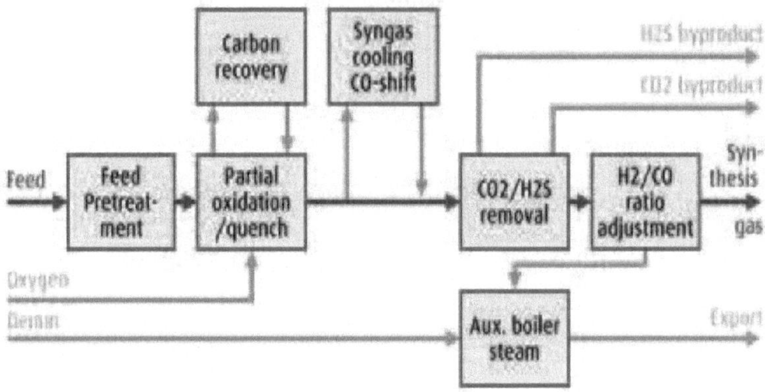

Figure 13: Process involving no catalytic converter (Linde, 2012).

The purification of hydrogen and carbon monoxide are both achieve with chemical and physical wash processes being used to remove the CO_2, and also a low-temperature processes for the production purification, which is done with a downstream pressure swing adsorption unit to produce high-purity hydrogen (figure 14). A typical plant capacity is also presented in table 3.

Figure 14: An example of a natural gas plant in Turkey, constitute a partial oxidation, CO2 Wash, and Membrane Separation, PSA. Star-up in 2003 (Linde, 2012)

Oxogas	60,000 Nm³/h
	53.8 mmscfd
H_2	28,000 Nm³/h
	25.1 mmscfd

Table 3: Typical plant capacity for a CCS plant illustrated in figure 14 above

Advantages of Pre-Combustion Capture

- It is a proven industrial scale technology in oil and gas refineries, but its needs 3x scale-up for power plants.
- Its captured up to about 90-95% of CO_2 emissions
- It is applicable to natural gas, and also coal fired IGCC power plants.

- It has a low technological risk, and may become the most efficient method.
- The process can produce H_2 as transportable energy vector, or liquid fuels from coal, but with penalties on efficience.

Disadvantages of Pre-Combustion Capture

- It requires a chemical plant in front of gas turbine
- It has a high investment cost of dedicated new build plant (it can't be retrofitted).
- It has a consequence of emitting high NOx and will require an expensive scrubber.
- The efficiency of H_2 burning turbines is lower than conventional turbines

2.4 Oxy-fuel Combustion Capture

Oxy-fuel combustion method involves burning of the fossil fuels in almost pure oxygen instead of air, which result in a more complete fuel combustion. The results in an exhaust stream, would consists of almost pure CO_2 (typically 90%) and water vapour, which can then be easily separated from the CO_2 by condensation. The low Nitrogen concentration of the flue gas when air is used for combustion enhanced the prospective application of the method, since further downstream treatment of CO_2 is not required. Steam conditions and initial quantities of oxygen required are similar to those set for previous technique but research show that the method is at early stage of practise in developed countries like Europe (Galadima and Garba, 2008).

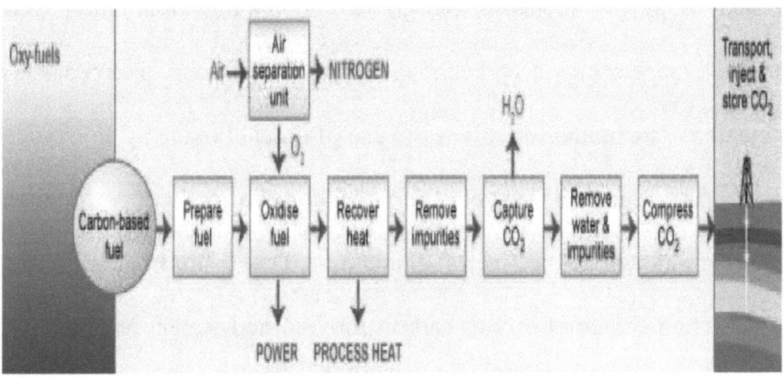

Figure 15: Process involving oxy-fuel combustion (Scottish, 2012).

The main problem attributed to this method is the consequence of separating oxygen from the air. The process is usually a complete cryogenically, that requires a lot of energy, for example, a typical 500MW coal-fired power station supplying pure oxygen requires at least

15% of the electricity the plant generates annually (Scottish, 2012). However, a promising new technology under development known as chemical looping will provide a process in which the oxygen in the air is removed by oxidation of a metallic compound which can then be reduced during combustion, allowing the oxygen to be released for the process.

2.5 The Chemical Looping Process

Chemical-looping combustion (CLC) is a technology that provides a means of separating a greenhouse gas CO_2 from hydrocarbons combustion. The method involves the use of a metal oxide as an oxygen carrier which transfers oxygen from the combustion air to the fuel, avoiding direct contact between fuel and air (Second Nordic, 2001). The products from the combustion, carbon dioxide and water, will then be kept separate from the rest of the flue gases, such as nitrogen and also any remaining oxygen. The method can be achieved in two forms; (Second Nordic, 2001)

i) Two reactors in the form of interconnected fluidized beds with a fuel reactor where the metal oxide is reduced by reaction with the fuel, and

ii) An air reactor where the reduced metal oxide from the fuel reactor is oxidized with air.

The outlet gas from the fuel reactor consists of CO_2 and H_2O, while the outlet gas stream from the air reactor contains only N_2 and some unused O_2 (figure 16). The net chemical reaction for both methods over the two reactors is the same as for normal combustion with the same amount of heat released. This process is in contrast to the other known methods for separating carbon dioxide from flue gas, in which a large amount of energy and expensive equipment are required for carrying out the separation. All parameters, important for the design of a CLC system are clear with respect to oxygen carrier reactivity and fluidization properties. In a chemical-looping combustion (CLC) system, which comprises of two reactors, an air and a fuel reactor, the process fuel is introduced to the fuel reactor in a gaseous form where it reacts with the oxygen carrier, (a metal oxide, MyOx)

$$(2n+m)MyOx + C_nH_2m \rightarrow (2n+m)MyOx\text{-}1 + mH_2O + nCO_2 \quad (1)$$

The evolve gas from the fuel reactor contains CO_2 and H_2O; this means that pure CO_2 can be captured when H_2O is condensed. The reduced metal oxide, MyOx-1, is then circulated to the air reactor where it is then oxidized and therefore regenerated:

$$MyOx\text{-}1 + \frac{1}{2}O_2 \rightarrow MyOx \quad (2)$$

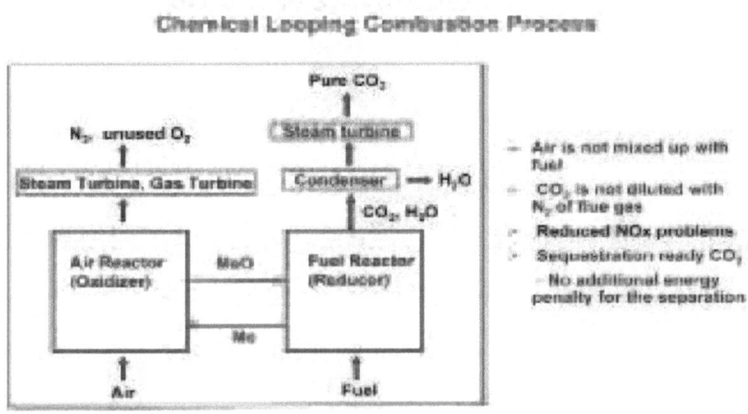

Figure 16: chemical looping combustion process (Carbon Sequestration, 2009).

The flue gas from the air reactor will contain N_2 and any unreacted O_2. Reaction (1) is often endothermic while reaction (2) is exothermic (Second Nordic, 2001). The total amount of heat evolved from reaction (1) plus (2) is the same as for normal combustion (Second Nordic,

2001). The advantage with this system is that the CO_2 and H_2O are inherently separated from the rest of the flue gases, and no energy is expended for this separation. This should be compared to other technologies for CO_2 separation which, when used in a power process with natural gas as the fuel, have a relative reduction in the plant efficiency of 12-19% (Lyngfelt, and Leckner, 1999).

2.6 References and Further Readings

Angela D, michele A, and marcella N, Carbon dioxide capture by amines: increasing the efficiency by amine structure modification Fuel Chemistry Division Preprints, 47(1), 54, 2002.

Aresta, M.; Ballivet-Tkatchenko, D.; Belli Dell'Amico, D.; Bonnet, M. C.; Boschi, D.; Calderazzo, F.; Faure, R.; Labella, L. and Marchetti, F. Chem. Commun. 2000, 1099.

Aresta, M.; Ballivet-Tkatchenko, D.; Bonnet, M. C.; Faure, R.; and Loiseleur, H. J. Am. Chem. Soc. 1985, 107, 2994

Carbon Sequestration: Oxygen Carrier Development for Chemical Looping Combustion; National Energy Technology Laboratory1450 Queen Avenue SW Albany, OR 97321-2198 541-967-5892 April, 2009

Davison, J., Freund, P., and Smith, A.: "Putting Carbon Back Into the Ground," IEA Greenhouse Gas R&D Programme (2001).

Dubey MK, Ziock H, Rueff G, Elliot S, Smith WS, Lackner KS, Johnston NA. Extraction of carbon dioxide from the atmosphere through engineered chemical sinkage. Abstracts of papers of the American Chemical Society 2002;233(1): U577-U578.

Galadima, A. and Garba,, Z.N (2008) Carbon capture and storage in Nigeria: Fundamental science and potential implementation risks. Science World Journal, Volume 3, Number 2, 95-99.

IEA 2006. World Energy Outlook 2006. Paris OECD/International Energy Agency (IEA); available from http://www.eia.org

Ikeme, J. 2008. Assessing the Future of Nigeria's Economy: Ignored Threats from Global Climate Change Debade. Africa Economic Analysis; available from http://www.afbis.com/analysis/climate_change.htm

IPCC 2007. "Climate Change 2007": Summary for policy makers. The AR4 Synthetic Report. Intergovernmental Panel on Climate Change: Released 17, Nov. 2007, Valencia, Spain.

Irons, R.; Sekkappan, G.; Panesar, R.; Gibbins, J. & Lucquiaud, M. 2007. CO_2 Capture Ready Plants. International Energy Agency [IEA], Technical study Report Number 2007/4.

Keeling, C.D., and Whorf, T.P.: "Atmospheric CO2 Records From Sites in the SIO Air Sampling Network." In Trends: A Compendium of Data on Global Change. Carbon Dioxide Information Analysis Center, Oak Ridge National Laboratory, U.S. Department of Energy, Oak Ridge, Tenn., U.S.A (2005). Available from: http://cdiac.esd.ornl.gov/trends/co2/sio-mlo.htm

Linde engineering: Synthesis Gas process accesses 2012, Available from:http://www.lindeengineering.com/en/process_plants/hydrogen_and_synthesis_gas_plants

Lyngfelt, A., Kronberger, B., Adanez, J., Morin, J., and Hurst, P.: "The Grace Project: Development of Oxygen Carrier Particles for Chemical-Looping Combustion, Design and Operation of a 10 Kw Chemical-Looping Combustor," GHGT-7 Conference, Vancouver B.C., Canada, September 5-9, 2004.

Lyngfelt, A. and B. Leckner. Technologies for CO_2 separation. In Mini-symposium on CO_2 capture and storage. 1999. Göteborg: Chalmers University of Technology andUniversity of Gothenburg

Malumfashi, G. I. 2007. Phase-out of gas flaring in Nigeria by 2008. The prospect of a multi-Win project (Review of the Regulatory, Environmental and Socio-Economic Issues). Swiss National Centre for Competence in Research (NCCR), Berne, Switzerland; Available from http://www.nccr.trade.org/

Post note 335, CO_2 capture, transport and storage, June 2002

Rosenzweig, C. & Casassa, G. 2007. Assessment of observed Changes and Responses in Natural and Managed Sytems. IPCC Forth Assessment Report (AR4). Chapter 1.

Scottish Carbon Capture & Storage (CCS) modified Jan 2012: Containment solutions to complement emissions reduction strategies. http://www.geos.ed.ac.uk/sccs

Second Nordic Mini-symposium on Carbon Dioxide Capture and Storage, Göteborg, Oct. 2001. Available at http://www.entek.chalmers.se/anly/symp/symp2001.html

Statistical Review of World Energy 2004, BP, Available from: http://www.bp.com/

CHAPTER THREE: CARBON (IV) OXIDE TRANSPORTATION

Transportation is a key part of current practice of carbon capture and storage where CO_2 is transported to sink. CO_2 transport is the necessary link between CO_2 capture, from point-of-capture to storage sites. The role focuses on safety, flexibility, reliability, and cost. Although CO_2 is transported via pipelines, ships, train, and tanker trucks for enhance oil recovery (EOR) and other industrial operations, with pipeline transport considered to be the most cost-effective and reliable method of transporting CO_2 for onshore CCS (Svensson et al., 2004).

The pipelines routine are used in carrying large volumes of natural gas, oil, condensate and water over distances of thousands of kilometres, both on land and in the sea. Pipelines are laid in deserts, mountain ranges, heavily populated areas, farmland and the open range, in the Arctic and sub-Arctic, and in seas and oceans up to 2200 m deep (IPCC, 2005). The present focus on safety, flexibility, reliability, and cost reductions has created a demand for new solutions that can compete with pipeline transport (IEA, 2002; Gale and Davidson, 2004). The various

CO_2 sources and the dramatically increase in the growth of the CO_2 market require a flexible and reliable transport system. However, little information on large-scale ship, train, and truck-base transport of CO_2 are known (Svensson et al. 2003; Yara, 2005). As a result smaller dedicated CO_2 ships are in operation today (Anthony, 2004), (Bock et al, 2003). The size of these ships is up to 1500 m^3, and the transport pressure is about 14-20 bar. Pressure and temperature are important given a prior attention in transportation of CO_2. Lower pressure is required for enlarged storage and ship tanks, which again requires new processes. Ship transport of CO_2 provides a more flexible, and in many cases, more cost- effective transport solution than pipelines.

Logistically, the efficiency of the CO_2 transportation is situated behind the CO_2 captures facilities and delivering a link to storage-sites. The selection of the transport system basically depends on three parameters: capacity, distance between source and sink, The transport infrastructure for CO_2 should be able to carry big quantities as in ships, trains and trucks and also to guarantee a continuous mass flow as in pipeline method (figure 17).

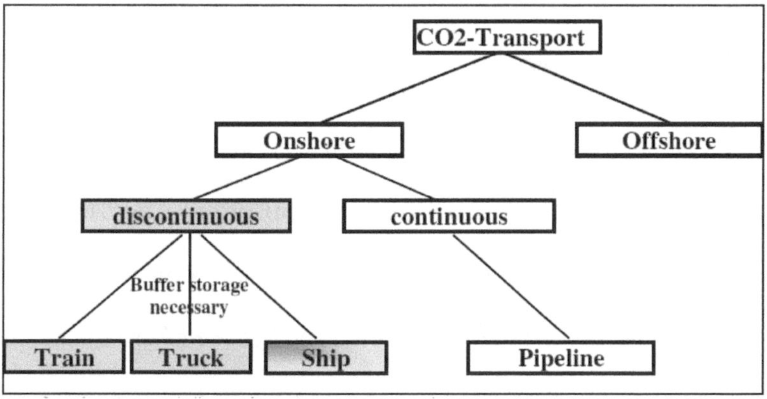

Figure 17: CO_2 transportation chart. Pipeline transport is a continuous while transportation via ships, trucks and rail is discontinuous. Although, the combination of both transport (continuous and discontinuous) is generally possible.

Carbon dioxide transport is currently the least complicated element in the CO_2 capture and storage chain as the technology is already in existence, adopting the method for transporting LPG and at same time costs can also be realistically estimated. The main challenges with CO_2 transport is that, CO_2 behaves differently under varying pressures and temperatures and therefore transport of CO_2 must be carefully controlled to prevent solidification and blockages occurring. Thus, has to undergo many processes before transportation.

3.1 Some Useful Factors

3.1.1 CO_2 Compression

Gas compression is well developed in the natural gas industry around the globe and it uses matured technologies. CO_2 compression uses the same equipment as natural gas compression, with some modifications to suit the properties of CO_2. Avoiding corrosion and hydrate formation are the main additional operating issues when dealing with CO_2. Compressors come in many different types (e.g. centrifugal, reciprocating and others), makes and sizes. Centrifugal compressors are usually the preferred type for large volume applications because of their ability to handle large flow rate (to hundred thousands of cubic feet per minute). In 2002, 490,000 hp of compression was installed in the USA, with a capital investment of over $ 635 million (True, 2003).

In 1998, more than 25 million tonnes of CO_2 were captured, compressed, transported and injected in the Permian basin in the USA to recover nearly 150,000 barrel/day of oil through CO_2 enhanced oil recovery (EOR) schemes (Stevens et al., 2000). This illustrates that in some areas of the world there is more than adequate operating experience in compressing and handling CO_2 in large-scale applications. However, as the oil reservoirs are maturing, for example, those in China and Indonesia, the prospect of injecting CO_2 to enhance the oil production or for storage can be technically and economically feasible. If these projects are to be implemented, large-scale handling of CO_2 is required.

The main purpose to compress CO_2 is to make it more efficient to transport. Certain amount is needed for the transport and can be calculated by using a phase diagram of CO_2 which can be seen in figure 18: the phase diagram is a pressure-temperature relationship. This shows boundaries of three phases:

- Solid phase,
- Liquid phase, and
- Gas phase.

There are two important points to note on the phase diagram,

- The triple point: - this point occurs when the pressure is 0.25MPa and at a temperature of -56° C. At this point, solid, liquid and gaseous phases co-exist together. CO_2 can only exist in either the gaseous or the solid phase below this temperature and pressure
- The critical point: - this occurs at a p pressure of 7.38 MPa and a temperature of 31.4°C. Above this critical pressure and at higher temperatures than –60°C, only one condition exists: that of the supercritical/dense phase.

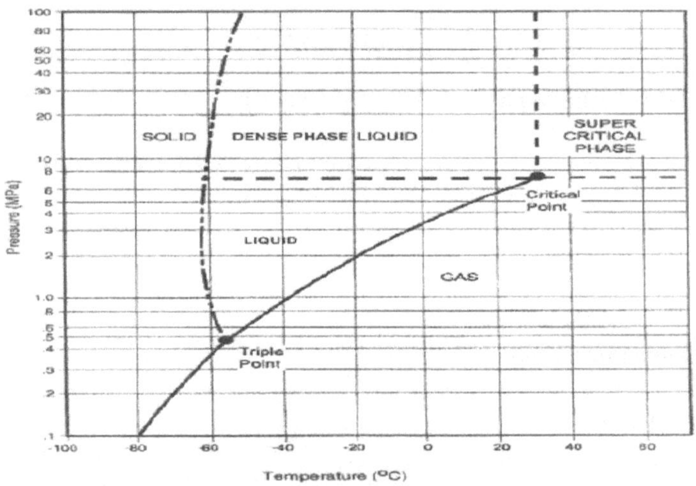

Figure 18: Pressure, temperature relationship phase diagram for CO_2 (Asia Pacific, 2004).

The following figure (i.e figure 19) shows that the density of CO_2 changes as the pressures and temperatures vary. Above the critical pressure of 7.38 MPa and at temperatures lower than 20°C, CO_2 would have a density between 800 to 1,200 kg/m^3. If this is compare to the density of water which is 1,000 kg/m^3, a higher density is favourable when transporting liquid CO_2, as it is easier to move a dense liquid than a gas. Therefore it is typical to compress CO_2 to above 7.38 MPa for efficient transportation.

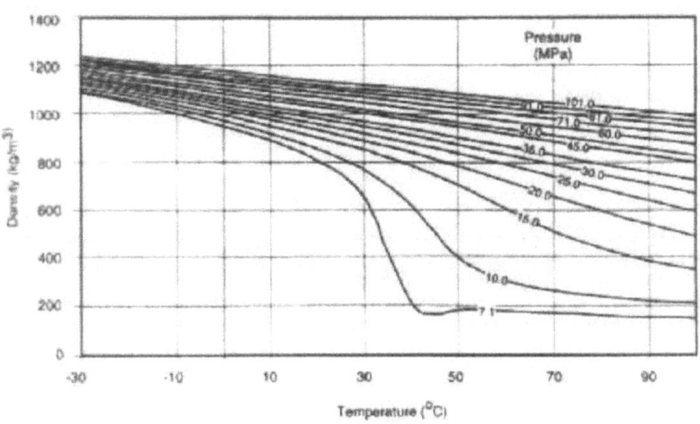

Figure 19: CO_2 density diagram (Bachu, 2003).

A number of stages of compression will be required before optimal pressure is achieved for transport of the CO_2. This is because, from engineering principle, it is impossible to raise the pressure of a gas such as CO_2 ten to twenty fold in one step, as this would result in too high a temperature rise in the gas. Therefore, compression generally occurs in a number of steps or stages. To determine the number of stages of compression, it is first necessary to select the acceptable compression ratio per stage. This ratio is generally in the order of 3 or 4. Table 1 shows the suction and discharge pressures and temperatures of a four-stage compressor with a compression ratio of 3. If needed, a fifth stage of compression can also be added. There is a considerable rise in the gas temperature during each stage as can be seen from the table. Aerial coolers are generally used to cool the process CO_2 stream to the appropriate suction temperature.

Stage No.	Suction Pressure kPa	Suction Temp. °C	Discharge Pressure kPa	Discharge Temp. °C
1	150	40	455	156
2	385	35	1160	149
3	1090	35	3270	150
4	3200	35	9600	152

Table 4: Suction and discharge pressure and temperature as complied through the computer model (HYSIS) simulation (Asia-Pacific, 2004).

3.1.2 Transport Composition of Carbon Dioxide

Captured CO_2 is conditioned to remove impurities and compressed for transportation. CO_2 is defined as a fluid consisting of more than 90 percent CO_2 molecules compressed to a supercritical state. Currently, there are no composition requirements (e.g., moisture or co-constituents) for the transport and geologic storage of CO_2 (MIT, 2007). And also no established standard for permitted levels of impurities in CO_2 for CCS, the pipeline-quality CO_2 compositions adhered to by the major EOR pipeline operators constitute best practice. Currently, these requirements are built into contracts between the supplier and the transporter and between the transporter and the end user.

The captured CO_2 may contain impurities like water vapour, H_2S, N_2, methane (CH_4), O_2, mercury, and hydrocarbons which may require specific handling or treatment before transport, the CO_2 has to be dry to levels of below 50 ppm of water. Presence of water above this level is not desirable from an operational standpoint (Aspelund and Jordal, 2007). CO_2 reacts with water to form carbonic acid, which is corrosive.

$$CO_2 + H_2O \rightarrow H_2CO_3 \rightarrow H^+ + HCO_3^- \rightarrow H^+ + CO_3^{2-}$$

Additionally, under the appropriate thermodynamic conditions, hydrates (solid ice-like crystals) can form and block the pipeline (Barrie et al., 2004).

H_2S is toxic, even at low concentrations of 200 ppm. Pipelines containing H_2S will require extra due diligence, particularly near populations. However, it is important to note that it is possible to safely store H_2S with CO_2. Developed facilities in Canada were known for disposing of H_2S, by injecting it in geologic formations since 1989 (Heinrich et al. 2004). Injection of acid gas also currently occurs at 39 active operations in Alberta and north eastern British Columbia (CSS Guideline, 2008). Economically, surface desulfurization through the Claus process is not feasible, thus, the surface storage of the produced sulphur constitutes a liability, and therefore more operators are turning to acid gas disposal by injection into deep geologic formations.

$$H_2S + 3/2O_2 \rightarrow SO_2 + 2H_2O$$

$$2H_2S + SO_2 \rightarrow 3S + H_2O$$

Since H_2S react with water to form sulphuric acid, dehydration will be particularly important in these cases, because the acid form is highly corrosive and may also result in pipeline cracking, increasing the some potential leaks.

$$H_2S + O_2 \rightarrow H_2O + SO_2$$

$$SO_2 + [O] \rightarrow SO_3$$

$$SO_3 + H_2O \rightarrow H_2SO_4$$

Reaction of H_2S in water:

$$H_2S \rightarrow HS^- + H^+ \quad \text{(formation of disulphide ion)}$$

$$HS^- \rightarrow S^{2-} + H^+ \quad \text{(formation of sulphide ion)}$$

$$H_2O \rightarrow H^+ + OH^- \quad \text{(self-ionization of water)}$$

Presence of CH_4 affects the exhibited vapor pressure of CO_2. This in consequence, complicates the accurate prediction of flow (Svensson,

2004). Particularly in EOR applications, where organic materials are present for bacteria, oxygen is tolerable only in minute quantities (10 ppm). Even in deep saline formations organics may be present, and significant quantities of oxygen in the gas stream could allow for formation of bacterial colonies, affecting the injection operations.

Mercury is present in coal and is a natural by-product of the combustion process. Thus, this could condense in the pipeline system and create operational complications as well as implications for storage. Co-constituents with CO_2 are unknown on a large scale and will most likely affect the requirements for CO_2 purity. For the sole purpose of storage, the threshold for impurities could be different; hence, the processing requirements and pipeline standards could be uniquely prescribed for a particular project

Due to the presence of this corrosive compound a strict control of the water content in the CO_2 stream is essential for safe and efficient operation of a compressor. A glycol dehydrator is often installed ahead of the CO_2 compression stream to control water content to an acceptable level. Glycol dehydration is a standard unit operation in the oil and gas industry. For smaller scale operation, the dehydration unit may be eliminated if the temperatures of the gas stream at the intercooler stages can be controlled to drop off the water. This technique is widely practiced in acid gas compression in Western Canada

3.1.3 Transport Quality Specification of CO_2

Although no permanent standards have been fully established, the following specifications are recommended for high quality CO_2 transportation.

a) Carbon Dioxide. Product shall contain at least ninety-five mole percent (95%) of Carbon Dioxide as measured at the SACROC delivery meter.

b) Water. Product shall contain no free water, and shall not contain more than 0.48 9 m-3 in the vapour phase.

c) Hydrogen Sulphide. Product shall not contain more than fifteen hundred (1500) parts per million, by weight, of hydrogen sulphide.

d) Total Sulphur. Product shall not contain more than fourteen hundred and fifty (1450) parts per million, by weight, of total sulphur.
e) Temperature. Product shall not exceed a temperature of 48.9 oC.
f) Nitrogen. Product shall not contain more than four mole percent (4%) of nitrogen.
g) Hydrocarbons. Product shall not contain more than five mole percent (5%) of hydrocarbons and the dew point of Product (with respect to such hydrocarbons) shall not exceed –28.9 oC.
h) Oxygen. Product shall not contain more than ten (10) parts per million, by weight, of oxygen.
i) Glycol. Product shall not contain more than 4 x 10-5 L m-3 of glycol and at no time shall such glycol be present in a liquid state at the pressure and temperature conditions of the pipeline

3.2 CO_2 Transportation

CO_2 can be transported by land via pipelines, motor carriers or railway, or by ocean via ships

3.2.1 Ocean Based Transportation

1. **CO_2 Ship Transportation**

Carbon dioxide is continuously captured at the plant on land, but the cycle of ship transport is discrete, and so a marine transportation system includes temporary storage on land and a loading facility. The large scale transportation of CO_2 by ship has been limited to the food and brewing industry to date. These industries transport some 100,000 tonnes of CO_2 per year which is much less than the amount which would need to be transported for commercial CO_2 storage. The capacity, service speed, number of ships and shipping schedule will be planned, taking into consideration, the capture rate of CO_2, transport distance, social and technical restrictions. However, the conditions required for CO_2 transport are similar to those of Liquefied Petroleum Gas (LPG) of which large quantities are transported by ship. As such LPG transport provides extensive information relevant to creating large scale CO_2 transport networks. Current CO_2 transport ships carry liquid CO_2 at temperatures of -25 to -3^{0}C and pressures of between 1.4 to 1.7 MPa. The typical capacity of a ship is 850 to 1,400 tonnes of CO_2 which is too small for the large quantity of CO_2 that would need to be transported for CO_2 storage. However, semi-ref type ships used for LPG transport, which cool LPG to -7^{0}C and pressurise it to 0.7 MPa could hold up to 24,000 tonnes of CO_2.

At the delivery point, what would happen depends on the CO_2 storage system. If the point of delivery is onshore, the CO_2 is directly unloaded from the ships into temporary storage tanks and if the point of delivery is offshore, as in the ocean storage option, the ships might unload to a platform, that is to a floating storage facility (similar to a floating production and storage facility routinely applied to offshore petroleum production), to a single-buoy mooring or directly to a storage system.

As transportation by ship does not allow a continuous flow of CO_2 from the industrial source to the storage sink some sort of interim storage facilities would be required (e.g. in harbours). Currently LPG is stored either in large steel tanks or in underground rock caverns, both of which could be used for intermediate CO_2 storage.

Figure 20: A scenario of CO_2 transportation by a ship (CCS Technology, 2007).

These ships are not suitable for large-scale ship-based transport of CO_2 due to composition of the CO_2. Thus, the ship must be constructed as pressure vessel, which makes costs very high. And yet, lower pressure is required for enlarged storage and ship tanks. However, this should not be a major problem, as tankers are currently used for shipping liquefied petroleum gas (LPG) and tankers similar to these could be used for CO_2 (figure 21). Generally, CO_2 is tranported simillar to how LPG is been transported and all due consideration is given in an apropriate to it based on the LPG transportation.

Figure 21: Constructed pressure vessel for low pressure; this is already used in tranporting LPG (CCS Technology, 2007).

3.2.2 Land Based Transportation

There three potential systems to be consider for land-based CO_2 transport, namely:

1. pipeline
2. tankers truck, and
3. Railway

1. Pipeline Transportation

Transport of CO_2 by pipeline is common in the United States where some 3000km of CO_2 pipelines exist (figure 6). These have been in used since the early1970s for transport of CO_2 for Enhanced Oil Recovery

(EOR) projects (SCCS, 2012). Pipelining is currently the most economical method of transporting large quantities of CO_2, and therefore the preferred option. In addition, there are some 700 km of lateral pipelines, which distribute CO_2 from the main lines. Altogether there are some 3,100 km of CO_2 pipelines worldwide, with a capacity of 44.7 million tonnes per year of CO_2 (SCCS, 2012).

Figure 22: a scenario of a 300k pipeline for CO_2 transportation in USA (SCCS, 2012).

All large scale pipelines are currently designed for supercritical/dense phase CO_2 transport, in order to avoid solidification of the CO_2 and allow it to be pumped as a liquid. Typically pressures are kept above 10 MPa by recompression stations at intervals along the pipelines.

Impurities such as methane, hydrogen sulphide, nitrogen and water also affect the properties of CO_2 as mentioned above. Water is the most critical of these as CO_2 dissolved in water creates an acidic solution that attacks metal pipelines. Additionally solid ice-like CO_2 hydrate crystals can form when CO_2 and water mix causing pipeline blockages and for these reasons the CO_2 must be dried before transportation.

There are existing designs and safety criteria to ensure safe and reliable transport of CO_2. Pipeline designers consider the pressure, temperature, and properties of the fluid; the elevation or slope of the terrain; dynamic effects, such as earthquakes, waves, currents, live and dead loads, and thermal expansion and contraction; and the relative movement of connected components (figure 23). The compressibility and density of CO_2 undergo significant nonlinear variation in normal pipeline operating conditions (within normal pipeline pressure and temperature ranges). Therefore, the design of CO_2 pipelines requires point-by-point estimation of fluid properties using computational models (MRCSP, 2005). For pipeline construction, selection of pipe diameter, wall thickness, material strength, and toughness depends on the transmissible fluid's temperature, pressure, composition, and flow rate. For example, fluid flow rates are lower in larger-diameter pipes. Lower fluid flow rates result in fewer pressure drops, allowing a pipeline designer to consider reducing the pressure requirements for CO_2 entering the pipeline, or reducing the number of compressors along the pipeline.

Figure 23: undergoing construction of pipeline for CO2 transportation (CCS Technology, 2007)

There is very little cost data on CO_2 pipelines so far; however, there is an abundance of cost data for natural gas pipelines. As reported in the Oil and Gas Journal, pipeline costs come from: material, labour, right-of-way access and miscellaneous. The average cost per mile for the 2002 data shows few clear-cut trends related to either length or geologic area. In general, however, the cost per mile within a given diameter indicates that the longer the pipeline, the lower the unit cost (per mile) for construction. And, lines built near populated areas tend to have higher unit costs. Additionally, road, highway, river or channel crossings and marshy or rocky terrain each strongly affects the pipeline construction costs (True, 2003). Therefore, it should be noted that a specific pipelines crossing difficult terrain could have construction costs 50% and higher than the average cost (Heddle et al., 2003). Estimated pipeline construction cost to be US $ 20,989/in/km. They further estimated that the operating and maintenance (O&M) costs would be US $ 3,100/km, independent of pipeline diameter. It should be noted that this O&M cost estimate does not account for pumping or its associated costs. These cost data can be used to estimate the construction costs for CO_2 pipeline as there is little difference between land construction costs for these two types of pipelines, except that the CO_2 pipelines might be slightly more expensive because of the greater wall thickness needed to contain the CO_2 (Heddle et al., 2003).

The advantage of pipeline transportation of CO_2 is that it can deliver a constant and steady supply of CO_2 without the need for intermediate storage along a distribution route. Ship transportation of large quantities of CO_2 may be feasible when it needs to be transported over long distances or overseas; however, many anthropogenic CO_2 sources are located far from navigable waterways, so such a scheme will still most likely require pipeline construction between CO_2 sources and port terminals and also most economical for transporting large quantities of CO_2 over land.

Pipeline	Location	Operator	Capacity ($MtCO_2\ yr^{-1}$)	Length (km)	Year finished	Origin of CO_2
Cortez	USA	Kinder Morgan	19.3	808	1984	McElmoDome
Sheep Mountain	USA	BP Amoco	9.5	660	-	Sheep Mountain
Bravo	USA	BP Amoco	7.3	350	1984	Bravo Dome
Canyon Reef Carriers	USA	Kinder Morgan	5.2	225	1972	Gasification plants
Val Verde	USA	Petrosource	2.5	130	1998	Val Verde Gas Plants
Bati Raman	Turkey	Turkish Petroleum	1.1	90	1983	Dodan Field
Weyburn	USA & Canada	North Dakota Gasification Co.	5	328	2000	Gasification Plant
Total			49.9	2591		

Table 5: Existing long distance CO_2 pipelines (Gale and Davison, 2002).

1. Tanker Trucks Transportation

The liquefied CO_2 can also be transported by motor tank trucks with trailers just like liquefied natural gas (figures 24 and 25). This can be stored in cryogenic vessels, having an inner vessels or “liquid container” which is surrounded and supported by an outer vessel or “vacuum jacket”. The space between the two is filled with a natural material that provides insulation. The tankers can be seen to have higher standards than ships in general, this is because most of the tanker incidents are led to public concern and can be controlled by careful navigation along prescribed routes, and by rigorous standards of operation. Although, LNG tankers are potentially dangerous, but are carefully designed and appear to be operated to very high standards.

Figure 24: CO_2 tanker trucks for oil field applications in China (Asia Pacific, 2004).

Carbon dioxide tankers and terminals are clearly much less at risk from fire, but there is an asphyxiation risk if collision should rupture a tank and the delivery system includes piping which carries gas from the vessel through the vacuum jacket to the outside, controlled by gauges and valves mounted outside the tank. These vessels are available in various sizes ranging from 2 to 30 tonnes, to suit customer's requirements. The conditions of the liquid CO_2 is typically at 1.7 MPa, -30°C. However, risk can be minimized by making certain that the high standards of construction and operation currently applied to LPG are also applied to carbon dioxide.

Figure 25: Refilling a CO_2 tanker truck (Asia pacific, 2004).

2. Railway Transportation

The railway system has a large carrying capacity that enables it to handle large volumes of bulk commodities over long distances (Asia Pacific, 2004). The captured CO_2 can therefore be transported in a specially and developed tank cars that are approved to transport liquid CO_2 at a pressure of 2.6 MPa. This is because, at this pressure, the net weight of liquid CO_2 that a single tank car can load is about 60 tonnes. It is known that CO_2 has been shipped by rail in two or three tank cars at a time, and that the tank cars can be left at the siding and serve as a storage tank until the next shipment arrives. However, there is no large scale CO_2 transport by railway at this point. Rail transport will only become a competitive transport option if the logistics can fit the volumes in the existing railway system. However, loading and unloading infrastructure and temporary CO_2 storage would also have to be included and would be costly.

Figure 26: A trail car tanker with potential for CO_2 transportation. (http://www.alaskarails.org/sf/other-cars/ME/ME-prox-41246.jpg)

One million tonnes of CO_2 per year is transported using existing rail infrastructure in Germany in a developed scenario to illustrate the potential of rail systems (Odenberger et al., 2003). The unit train capacity is 1,200 tonnes. Therefore each unit train will have 20 cars, each with a net weight of 60 tonnes. The loading and unloading time is assumed to be 4 hours and the total utilization is 7,500 hours per year (Asia pacific 2004). It is also determined that it would cost 5.5 Euro per tonne of CO_2 for the 250 km distance (i.e. about $ 2.2 /tonne/100 km). This cost did not include any loading and unloading infrastructure or temporary CO_2 storage, meaning that the actual cost would be higher. Rail lines do not extend to CO_2 source and storage site, and therefore fitting these sites with existing rail system is logistically complicated and may be problematic. Thus if a new railway line had to be laid, the cost would be even higher.

However, studies indicate that this system can only be suitable for very small scale application (GF, 2008). Another major challenge is that up to now storage has not been practised offshore and most suitable phase conditions are not fully evaluated.

3.3 References and Further Readings

Anthony Veder Group. 2004. URL. Available at: www.anthonyveder.nl

Aspelund, A., and K. Jordal. July 2007. "Gas Conditioning". The Interface Between CO_2 Capture and Transport." International Journal of Greenhouse Gas Control 1(3):343-354.

Asia pacific economic co-operation: Building capacity for CO2 capture and storage in the APEC region, 2004. Available at www.ccs101.ca/building_capacity_forccs

Bachu, Stefan. June 2003. "Variation of CO_2 Density as a Function of Temperature and Pressure." In "Screening and Ranking of Sedimentary Basins for Sequestration of CO_2 in Geological Media in Response to Climate Change." Environmental Geology 44(3):277–289

Barrie J., K. Brown, P.R. Hatcher, and H.U. Schellhase. 2004. "Carbon Dioxide Pipelines: A Preliminary Review of Design and Risks." In Proceedings of the 7th International Conference on Greenhouse Gas Control Technologies. September 5–9, Vancouver, Canada. http://uregina.ca/ghgt7/PDF/papers/peer/126.pdf

Bock, B., et al., Economic Evaluation of CO_2 Storage and Sink Enhancement Options. 2003, TVA Public Power Institute: Muscle Shoals, AL

CCS Technology: Capture, transportation and storage of CO2, Brendan Beck, IEAGHG, CCS-com Africa 2007: Available at: www.ieagreen.org.uk

CCS Guidelines: Guidelines for Carbon Dioxide Capture, Transport, and Storage. World Resources Institute (WRI). Washington, DC: WRI., World Resources Institute 10 G Street, NE Suite 800. Washington, DC 20002. 2008: available at: htpp://pdf.wri.org/ccs_guidelines.pdf

Gale, J.J. and Davison, J.E., 2002 6th International Conference on Greenhouse Gas Control Technologies (GHGT-6) Kyoto, Japan, October 2002. Transmission of CO_2 – safety and economic considerations. Elsevier Publishers

GF 2008. How can CO_2 be transported once it is capturerd?. Green Facts (GF). Facts on Health and the Environment. Available at: http://www.greenfacts.org/en/CO2_capture_storage/

Group III of the Intergovernmental Panel on Climate Change. http://www.ipcc.ch/pdf/specialreports/srccs/srccs_summaryforpolicymakers.pdf

Heddle, G., Herzog H. and Klett, M., The economics of CO_2 storage, Laboratory of Energy and the Environment, Massachusetts Institute of Technology, Cambridge, MA, USA 111p available at: http:// lfee.mit.edu/publications/ Publication No. LFEE 2003-003 RP, 2003

Heinrich, J.J, H.J. Herzog and D.M. Reiner. March 2004. Environmental Assessment of Geologic Storage of CO_2. MIT LFEE 2003-002. Prepared for the Clean Air Task Force. http://sequestration.mit.edu/pdf/LFEE_2003-002_RP.pdf

Hendriks, C., et al., 2002. IEA Greenhouse Gas R&D Programme. Building the cost curves for CO2
storage, Part 1: Sources of CO_2 PH4/9.

IEA Greenhouse Gas R&D Programme. 2003. Barriers to overcome in implementation of CO_2 capture and storage (2): Rules and standards for the transmission and storage of CO_2 PH4/23.

IPCC Special Report: Carbon Dioxide Capture and Storage: Summary for Policymakers 2005. A Special Report of Working

MIT 2007—Massachusetts Institute of Technology. 2007. The Future of Coal: Options for a Carbon-Constrained World. Cambridge, MA. Available at http://web.mit.edu/coal/.

MRCSP 2005—Midwest Regional Carbon Sequestration Partnership. December 2005. Midwestern Regional Carbon Sequestration Partnership Phase 1 Report. Available at: http://216.109.210.162/.

Odenberger, M. and Svensson R., Transportation systems for CO_2 – Application to carbon sequestration, Chalmers University of Technology, Goteborg, Sweden, 48p. available at: http://www.entek.chalmers.se/~klon/msc. 2003.

Stevens, S. and Gale J., Geologic CO_2 sequestration, Oil and Gas Journal, p40-44, May 15, 2000.

Svensson, R., Odenberger, M., Johnsson, F. and Strömberg, L., 2003. Transportation systems for CO_2: Application to carbon capture and storage. Energy Convers. Mgmt, Elsevier Science Ltd.

SCCS, Scottish carbon capture and storage: available at: www.goes.ed.ac.uk/sccs/ Date: 2012.

Svensson, R., et al., Transportation systems for CO_2-application to carbon capture and storage: Energy Conversion & Management, 2004. **45**: p. 2343-2353.

True, W., US pipeline companies solidly profitable in 2002, scale back construction plans, Oil and Gas Journal, p 60-90, September 8, 2003.

U.S. DOT/STB—U.S. Department of Transportation, Surface Transportation Board. Available at: http://www.stb.dot.gov/

Yara (former Norsk Hydro Agri). 2004. Available at : www.yara.com.

http://www.alaskarails.org/sf/other-cars/ME/ME-prox-41246.jpg

CHAPTER FOUR: GEOLOGICAL CO_2 STORAGE

After successful capture and transportation of carbon (IV) oxide as discussed in the previous chapters, there is need for it to be adequately stored, for it to be isolated completely from the atmosphere.

Herzog (2001) identified the following key factors that are necessary for an effective carbon dioxide storage option.

-The storage should be for a very long period of time, preferably 1000 years.

-The costs of transport and storage should be minimised

-All potential risks and accidents should be identified and eliminated

-Environmental impacts should be taken into account and

-The storage option should be in accordance with all national and international laws and regulations.

Various forms have been conceived for permanent storage of CO_2, among them include gaseous storage in various deep geological formations (including saline formations and exhausted gas fields), liquid storage in the ocean and solid storage by reaction of CO_2 with metal oxides to produce stable carbonates. This chapter will discuss the different geological formations option of carbon (IV) oxide storage while the next chapter will discuss the other two options.

4.1 Geologic Storage of Carbon Dioxide

Geologic storage also known as geo-sequestration of carbon dioxide (CO_2) involves storage of the greenhouse gas in underground formations, after it has been captured from power plants or other large industrial facilities. The carbon dioxide (in supercritical form) is injected into underground geological formations. Oil fields, saline formations, unmineable coal seams and saline-filled basal formations have been suggested as storage sites. Various physical and geothermal trapping mechanisms will prevent the carbon dioxide from escaping to the surface. It is an idea that is being pursued around the world in view of its potential to deliver significant reductions in CO_2 emissions. As part of a broader portfolio of technologies, geologic sequestration appears to be capable of playing an important role in stabilizing CO_2 concentrations in the atmosphere.

All options for reducing carbon dioxide emissions can be evaluated by studying the potential of geologic reservoirs and geologic processes to store as shown in the figure below (figure 27).

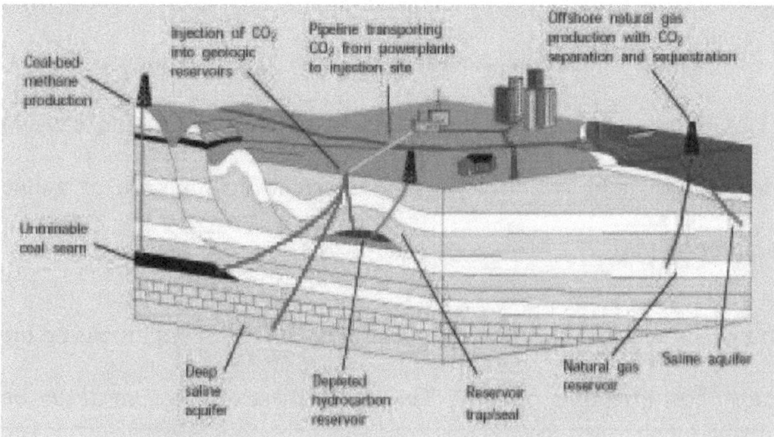

Figure 27. Potential CO_2 sequestration reservoirs and products. Source: (http://pubs.usgs.gov/fs/fs026-03/fs026-03.html).

The concept of controlling the level of CO_2 in the atmosphere is generally referred to as "carbon management." Carbon management includes the following strategies to reduce carbon emissions: carbon capture and sequestration, enhanced efficiency of power use and generation, the use of low-carbon fuels, and the use of renewable energy sources.

Carbon sequestration, broadly defined, is a term that includes the removal of CO_2 from the atmosphere by agricultural modifications and reforestation as well as the reduction of CO_2 emissions by capture and storage. Storage of anthropogenic CO_2 within geologic reservoirs is a method of carbon sequestration. The U.S. Geological Survey (USGS) is currently studying geologic options for storing CO_2 in depleted oil and gas reservoirs, unmineable coal seams, or saline aquifers (figure 27).

Initial strategies for geologic sequestration of CO_2 should focus on the most cost-effective options. Therefore, these options need to be identified. Storage sites that are close to CO_2 sources (power plants) lead to relatively lower transportation costs. Furthermore, CO_2 sequestration in some oil, gas, and coal-bed-methane reservoirs has the potential to increase the amount of fossil fuel that can be removed from those reservoirs. These enhanced recovery methods -- for example, enhanced oil recovery (EOR), enhanced coal-bed-methane recovery, or enhanced gas recovery (EGR) -- have an economic payback that will help to subsidize the net cost of CO_2 storage. Additionally, one of the important goals of this project is to determine storage capacity of both enhanced recovery reservoirs and non-economic reservoirs.

The IPCC estimates that there is technical potential for at least 2,000 Gt CO_2 of storage capacity in geological formations worldwide, and each storage option is currently being explored through industrial and /or pilot scale projects. Key geological characteristics must be considered to determine the most suitable storage structures in any given geological setting (see table 6).

Geological Characteristic	**Comment**
Geology – what rocks?	Sedimentary rocks: sandstones, coal, salt, limestones? Igneous, metamorphic generally inappropriate for storage?
Reservoir/ Seal Pair	***Reservoir***: Permeable rocks with primary or secondary porosity to allow gas to fill pore spaces ***Seal:*** Overlying impermeable formation which effectively 'seals off' the underlying host rock, to prevent upward migration of gas
Structure	Basin style, aquifer (hydrocarbon or non-hydrocarbon bearing), anticlinal folds, faults, thrusts.
Potential for	Producing/ depleting/ non-existent

Hydrocarbons	
Depth	Critical for CO_2 to reach a dense, supercritical fluid phase, generally below 650m at standard temperature and pressures, to achieve maximum volume storage capacity.
Hydrogeology: **Porosity/ Permeability**	***Porosity***: permissive primary or secondary pore space between grains of the rock. May be occluded by secondary re-crystallisation of minerals such as silica, illite, dolomite, calcite etc, but may be enhanced by e.g dolomitisation. ***Permeability:*** connected permissive fluid/ gas pathways within the host rocks, may be enhanced by micro-fracturing, joints, faulting, folding and karstification.
Containment	Basin boundaries - are they faulted/ pinch-outs/ fold closures? Effectiveness of seal pair?
Compartmentalisation	Do faults cause the host reservoir to be compartmentalised? If so, multiple injection wells may have to be considered.
Pressure/ Temperature of Reservoir	P/T will influence the solubility, density and buoyancy of CO_2 with both generally increasing with depth
Geothermal gradient	Relatively warm and cool basins will behave in a different fashion depending on P/T conditions
Tectonic setting	The Irish continental shelf is located on an Atlantic Passive Margin and is relatively tectonically stable for the foreseeable future. The

	Irish Sea is weakly seismically active.
Potential for CO_2: wall rock interaction	CO_2 reacts with wall rock in the host reservoir albeit in a generally predictable fashion. However at depth, natural secondary mineral precipitation will occur and may cause occlusion of the reservoir porosity, particularly close to the injection site.
Reservoir Recharge: **Depletion Drive or** **Water Drive**	Post production, is the natural recharge of available pore space in a Reservoir by water or low pressure natural gas? If with water; the reservoir is said to have 'water drive'; if by gas, the reservoir is said to have 'depletion drive'. The latter is more favourable for CO_2 injection, as the pore space can be refilled to natural levels without over-pressuring, thus reducing the risk of leakage.

Table 6: Geological features to be considered. Source: IPCC Special Report on CCS 2005.

The IEA have summarized the requisite characteristics (table 7), which are largely determined by accessibility to captured volumes of CO_2, the storage capacity of the geological structure, the amounts and rates which can be injected and finally the security of the site.

Accessibility	Location economically accessible to the CO_2 source Operator has legal rights to storage at that site
Capacity	Formation/ structure has adequate porosity and permeability to store CO_2 Storage volumes are adequate
Injectivity	Formation/ structure can store CO2 at the rate required to serve the intended source(s)
Security	Well defined trapping mechanism(s) Sufficient depth to retain supercritical CO_2 Cap rock is impermeable, continuous and thick enough to prevent upward migration Geological environment is sufficiently stable to ensure integrity of storage site No pathway faults or uncapped wells penetrate the cap rock and storage formation

Table 7: Requisite characteristics to be considered. Source: IEA (January, 2008)

4.2 Carbon Dioxide (CO_2) Geological Storage Options

The four geological storage options are explained in a more detail as follows:

4.2.1 Use of CO_2 in enhanced oil (EOR) and gas (EGR) recovery

In enhanced oil recovery (EOR) a combination of CO_2 and water is pumped into depleted oil wells to re-pressurize wells and "push" additional oil toward production equipment. Where denser CO_2 is injected into the reservoir, it pushes less dense methane gas or immiscible oil upwards and thereby increasing the effective gas volumes which can be removed economically. The injected CO_2 also prevents de-pressuring of the reservoir as increased volumes of working gas are recoverable.

The Weyburn CO_2 enhanced recovery project in Saskatchewan uses about 1 million metric tons of CO_2 per year. Enhanced oil and gas recovery offers the highest near term potential for storing CO_2, as well as an opportunity to sequester carbon at low cost, due to the revenues from recovered oil or gas. Currently, the scope is limited economically to point sources of CO_2 emissions that are near an oil or gas reservoir.

In an enhanced oil recovery application, the fate of the CO_2 that remains in the reservoir is well understood. In these applications, however, CO_2 was considered an added production expense and the amount left in the reservoir was minimized. DOE currently is supporting research on the long-term monitoring and verification of CO_2 storage as well as additional information such as the optimum combination of sequestration and enhanced oil recovery.

4.2.2. Depleted Oil and Gas Reservoirs

Depleted oil and gas reservoirs works on the same principle as enhance oil recovery (EOR); whereby injected CO_2 replaces the extracted working gas, and maintains reservoir pressure. Care has to be taken not to over-pressure the reservoir, which could lead to increased buoyancy and upward movement of the injected CO_2, as well as potential seismicity in extreme cases.

In addition to enhanced recovery operations, depleted oil and gas fields are natural reservoirs for storing CO_2. Depleted reservoirs can make attractive CO_2 sequestration targets since they have already proven their ability to contain oil, gas and water for millions of years and their geologic character is well defined by previous exploration efforts.

4.2.3. Deep Saline Formations

In the case of deep saline formations, the existing water in pore spaces must be displaced to create appropriate volumes for injected CO_2 storage, while care must be taken not to overpressure the reservoir.

Deep saline reservoirs are saltwater formations located many thousands of feet below the earth's surface. These reservoirs have two important benefits as CO_2 storage options. First, the estimated carbon storage capacity of saline formations in the U.S. is very large, making them a viable long-term solution. And secondly, many existing large CO_2 point sources are within easy access of a saline formation injection point. As with all of these geologic formations, understanding how CO_2 moves within the formation and ensuring it stays there are key aspects of sequestration research.

National Energy Technology Laboratory (NETL) has initiated a number of field tests to study the behaviour of CO_2 after it is injected into a formation. NETL also seeks to characterize potential sites to ensure their suitability of the geology (earth and rock structures). Factors to confirm include:

• The impermeability of the rock (cap rock) above the proposed storage area to prevent CO_2 from gradually moving upwards in the formation. For example, the presence of thick shale without interconnected cracks would indicate an effective cap (shales have a texture that does not permit fluids or gases to move through them)

• The storage capacity of the rock formation, i.e., whether it can hold enough CO_2 to be worth the cost of injection

• The distribution of the CO_2 in the reservoir and the chemical reactions that occur between the CO_2 and the reservoir rock and fluids

• A lack of faults in the area of injection operation that would avoid migration of fluids.

4.2.4. Unmineable Coal seams for Enhanced Coal bed Methane Recovery (ECBM)

Enhanced coalbed methane recovery results from unmineable coal, where CO_2 adsorbs to the surface of coal as methane (CH_4) is extracted. With increasing pressure CO_2 adsorbs into the coal, but if pressures and temperature increase significantly, the coal can become plastic, thus decreasing its permeability.

Figure 28: Coal bed methane test well at Sealand, Chester, UK.

Coal beds in the subsurface typically contain large volumes of methane-rich natural gas. A high percentage of this gas is adsorbed naturally on to the surface of the coal. Currently, methods for recovering coal-bed methane involve depressurizing the reservoir by pumping water out of the coal, thus permitting those methane molecules adsorbed on the surface of the coal to be released as free gas. An alternative approach for methane recovery is to inject CO_2 into the coal bed. Experiments indicate that coal beds have an affinity to absorb approximately twice as many CO_2 molecules as compared to methane. Thus, the potential exists to displace and recover coal bed methane efficiently. CO_2 recovery of coal bed methane has been demonstrated in limited field tests; however, additional research is needed to understand and optimize the process.

Similar to the by-product value gained from enhanced oil recovery, the recovered methane provides a value-added revenue stream to the carbon sequestration process, creating a lower net-cost option. The US Department of Energy Office of Fossil Energy estimates that 90 percent of coal reserves in the U.S. are unmineable due to inadequate coal thickness, extreme depth or a lack of internal continuity of the coal bed. A substantial portion of these unmineable resources may be amenable to methane recovery from CO_2 injection.

Depleted oil and gas reservoirs are the most appropriate land storage option in the current situation because of their good potentials to store highly pressurized fluids for very long period of time (Herzog *et al.*, 1997). Some studies revealed that the quantity of CO_2 that can be used for EOR and related applications is low compared to the total emissions level. Trillion cubic feet of coal bed methane are available worldwide. In U.S A. Alone there is over 645 trillion cubic feet. For this reason, enhance methane recovery from coal bed would be a good future option. Deep saline formations are heterogeneous structures in the sub-surface, with varying porosity and permeability distributions. If CO_2 is injected into these formations, the increased perplexing in the formations and reduced reservoir permeability may inhibit the potential leak of the gas (Flett *et al.,* 2004).These structures will therefore provide a good storage option for very long time, more importantly if the formations can react with the gas to form carbonate rocks like dolomite

4.3. Merits and Demerits of Different geological formations as Carbon dioxide storage option

4.3.1. Use of Co_2 in EOR and EGR and Depleted Oil and Gas Reservoirs

CO_2 is sometimes injected into declining or depleted oil fields to enhance oil recovery. Approximately 30 to 50 million metric tonnes of CO_2 are injected annually in the United States into depleted oil fields. This option is attractive because the geology of hydrocarbon reservoirs is generally well understood and storage cost may be partly offset by the sale of additional oil that is recovered.

Disadvantages of old oil fields are their geographic distribution and their limited capacity, as well as the fact that subsequent burning of the additional oil so recovered will offset much or all of the reduction in CO_2 emissions.

4.3.2. Deep Saline Formation

The main advantage of saline aquifers is their large potential storage volume and their common occurrence while their major disadvantage is that relatively little is known about them especially compared to oil fields.

4.3.3. Unmineable Coal seams for Enhanced Coal bed Methane Recovery (ECBM)

Unmineable coal seams can be used to store CO_2 because the CO_2 molecules attach to the surface of coal. The technical feasibility however depends on the permeability of the coal bed. In the process of absorption the coal releases previously absorbed methane and the methane can be recovered (enhanced coal bed methane recovery). The sale of the methane can be used to offset a portion of the cost of the CO_2 storage. But one of the disadvantages is that burning the resultant methane will produce CO_2 which would negate some of the benefit of storing or sequestering the original CO_2.

In general, IPCC estimates that CO_2 could be trapped for millions of years if the geological storage sites are well selected, designed and managed and that the sites are likely to retain over 99% of the injected CO_2 for over 1000 years. In 2009, scientists were reported to have mapped 6000 square miles of rock formations in the U.S. that could be used to store 500 years of worth of U.S. CO_2 emissions.

4.4 References and Further Readings

http://www.netl.doe.gov/technologies/carbon_seq/index.html. The site includes a wide variety of information, including links to newsletters, journals, papers, news reports, a video and program plans. For a comprehensive overview of current activities and plans, link to the Carbon Sequestration Technology Roadmap and Program Plan.

http://www.netl.doe.gov/technologies/carbon_seq/refshelf/refshelf.html. See also, the web site of the MIDCARB research consortium (Midcontinent Interactive Digital Carbon Atlas and Relational DataBase) which presents the results of its research evaluating the capacity for geologic sequestration in the member States Illinois, Indiana, Kansas, Kentucky and Ohio at: http://www.MIDCARB.org.

The International Energy Agency's (IEA, 2008) website at: www.ieagreen.org.uk/ and companion site focusing on carbon capture and storage at: http://www.co2captureandstorage.info/.

The Intergovernmental Panel on Climate Change (IPCC, 2011): http://www.ipcc.ch/.

EIA 2007. Nigeria Energy Data: EIA International Short Term Energy Outlook, Tables 3a and 3b: http://www.eia.doe.gov/ [Accessed 22-07-2000].

Flett, M. A.; Gurton, R.M. & Taggart, I. J. 2004. Heterogeneous Saline Formations: Long-Term Benefits for geo-sequestration of greenhouse gases. GHGT7: Proceedings of the 7th International conference on Greenhouse Gas Technologies, 5-9 September, 2004, Vancouver, Canada.

Herzog, H.; Drake, E. & Adams, E. 1997. CO_2 Capture, Reuse, and Storage Technologies for Mitigating Global Climate Change. A white Paper, Final Report. DOE Number:DE-AF22-96PC01257.

Falkowski, P., Scholes, R.J., Boyle, E., Canadell, J., Canfield, D., Elser, J., Gruber, N., Hibbard, K., Högberg, P., Linder, S., Mackenzie, F.T., Moore, B., III, Pedersen, T., Rosenthal, Y., Seitzinger, S.,

Smetacek, V., and Steffen, W., 2000, The global carbon cycle; A test of our knowledge of Earth as a system: Science, v. 290, p. 291-296. (Also available online at http://gcte.org/Glob-C-cycle.pdf)

The Scottish Centre for Carbon Storage has developed a free interactive resource which locates proposed CCS sites worldwide and details basic project information – see www.geos.ed.ac.uk/sccs

.

CHAPTER FIVE: OTHER STORAGE OPTIONS

This chapter will focus on other carbon dioxide storage options which include ocean and mineral storage.

5.1. Ocean Storage

The combustion of fossil fuels emits an estimated 23 giga tonnes of CO_2 annually into the earth's atmosphere. If captured and stored, this would significantly benefit the global environment. However, a very large store would be required. The figure below (i.e figure 30) shows some scenarios of CO_2 concentrations which suggest major changes in climate (Houghton et al., 1985).

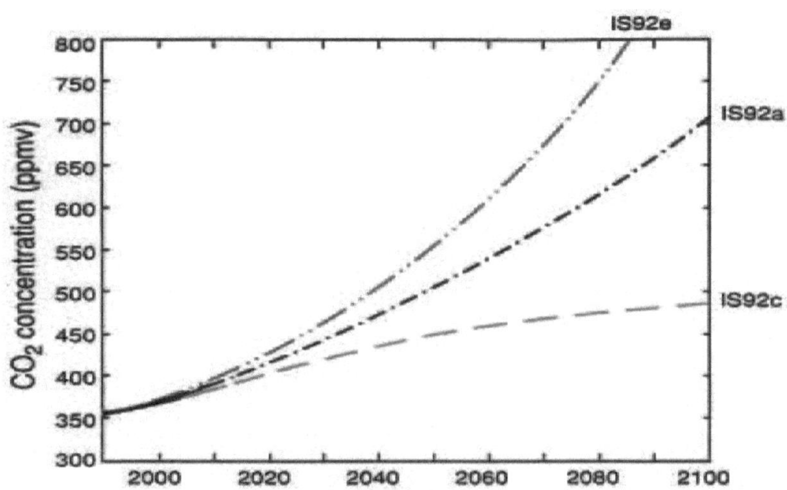

Figure 30: Some scenarios of CO_2 concentrations which suggest major changes in climate. Source: IEA greenhouse gas Research and Development Program.

Several types of potential reservoir are being considered, including the deep ocean, which is where CO_2 from the atmosphere would eventually be stored if left to natural processes. It has been estimated that the capacity of the deep ocean for deliberate storage of CO_2 may be more than 4000 Gtonnes. The figure below shows the concept of CO_2 storage in the ocean (Houghton et al., 1985).

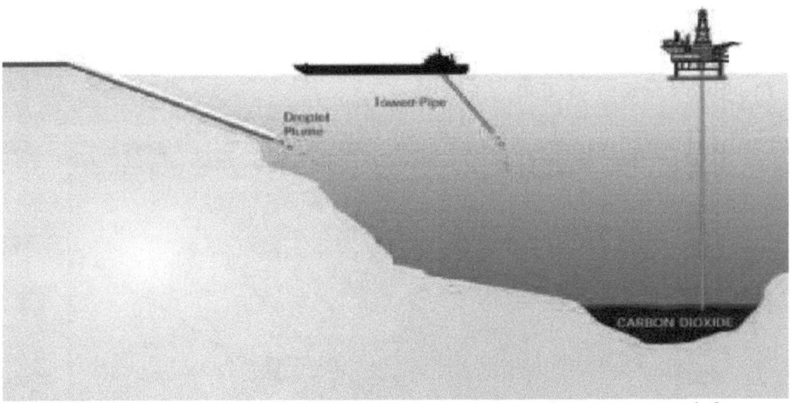

Figure 31: concept of CO_2 storage in the ocean. Source: IEA greenhouse gas research & development programme.

Oceans can accommodate CO_2 due to its appreciable solubility in water. Captured CO_2 could potentially be injected directly into deep oceans and most of it would remain there for millions of years (IPCC,

2005). Herzog (2001) reported that the ocean represents the largest storage option for the gas. Already 40,000 Giga tonnes of CO_2 are contained in the oceans compared to the atmospheric concentration of only 750 Giga tonnes. Thus, quantities of CO_2 that would double its atmospheric concentration would only change the ocean's concentration by 2% (Herzog, 2001). As such countries that are closed to the oceans have better chance of storing large volume of the gas. However, marine organisms are at serious risks due to resulting ocean acidity.

Several concepts have been proposed for the storage of CO_2 in the oceans which include:

- Dispersal of CO_2 in a very dilute form by injecting it using ship or pipeline into the water column at depths of 1000m to 2000m
- Depositing CO_2 directly onto the sea floor at depths greater than 3000m where it is denser than water and is expected to form a lake of liquid CO_2 on the seabed that would delay its dissolution into the environment.
- Formation of a sinking plume to carry most of the CO_2 into deep water.

- Storing the CO_2 in solid clathrate hydrate already existing on the ocean floor or growing more solid clathrate.

Of these options, the first is thought to be the most promising in the short-term. The cost of capturing the CO_2, transporting it 500 km and storing it has been estimated at $70/tonne CO_2.

According to IEA, before such a scheme could be employed, further research would be required to assess the potential benefits, advantages and disadvantages. For example, how much CO_2 could be stored and for how long? What would be the overall effect of such reductions in emissions? What adverse effects might occur? What changes would take place in the ocean, even without storage, if emissions continue as at present? In order to understand the importance of these factors, it is necessary to:

- Make reliable predictions of the technical feasibility and storage times.
- Understand how to predict and minimize any environmental impact.
- Make reliable cost estimates.

- Assess the net benefit of storing CO_2 in the ocean and compare this with other options.

This is not possible at present; specifically designed research programmes are needed to provide this information. Better understanding is also needed of the various natural processes, which determine the carbon cycle in the ocean.

Questions regarding the legal basis for such action (transportation of CO_2 in a similar way to LPG tanker) need to be addressed. For example, the discharge of CO_2 from ships has not been specifically considered for inclusion under the London Convention (which regulates discharges into the ocean), so at present it would not be permitted. In addition, the jurisdiction of governments wishing to use this option would have to be considered. Public understanding and opinion about the use of the oceans for this purpose will also be important.

5.1.1 Merits and Demerits of Ocean as CO_2 Storage Option

According to IEA, ocean storage has the potential to significantly reduce the emission of CO_2 to the atmosphere. It is one of the few techniques that might be able to handle the huge quantities of CO_2 involved but the environmental effects of oceanic storage are generally negative and poorly understood.

Large concentrations of CO_2 could kill ocean organisms; similarly the dissolved CO_2 would eventually equilibrate with the atmosphere so the storage will not be permanent.

Another problem is that water reacts with carbon dioxide to from carbonic acid according to the following equation.

$$H_2O + CO_2 \longrightarrow H_2CO_3,$$

Thus, increasing the acidity of the ocean water will negatively affects the aquatic organisms. Bicarbonate approach would reduce the pH effects and enhance the retention of CO_2 in the ocean but this would increase the cost and other environmental effects.

Another problem of ocean storage is the cost. It has been reported that the cost for deep ocean disposal of liquid CO_2 are estimated at about \$40-80/tonne of CO_2 (2002 USD). This figure covers the cost of sequestration at the power plant and naval transport to the disposal site.

5.2 Mineral Storage

In this process, CO_2 exothermically reacts with available metal oxides, which in turn produces stable carbonates. This process occurs naturally over many years and is responsible for a great amount of surface limestone. IPCC estimates that a power plant equipped with CCS using mineral storage will need 60-80% more energy than a power plant without CCS.

Carbon sequestration can be done by reacting naturally occurring Mg and Ca containing minerals with CO_2 to form carbonates as in the following equations

$$2Mg + O_2 \longrightarrow 2MgO, \qquad MgO + CO_2 \longrightarrow MgCO_3$$

$$2Ca + O_2 \longrightarrow 2CaO, \qquad CaO + CO_2 \longrightarrow CaCO_3$$

Most notable is the fact that carbonates have a lower energy state than CO_2 which is why mineral carbonation is thermodynamically favourable and occurs naturally. The raw materials such as magnesium based minerals are abundant and the produced carbonates are unarguably stable, so re-release of CO_2 into the atmosphere will not be an issue. The following table lists principal metal oxides of earth crust; theoretically up to 22% of this mineral mass is able to form carbonates.

Earthen Oxide	Percent of crust	Carbonate	Enthalpy change (kJ/mol)
SiO_2	59.71		
Al_2O_3	15.41		
CaO	4.90	$CaCO_3$	-179
MgO	4.36	$MgCO_3$	-117
Na_2O	3.55	Na_2CO_3	
FeO	3.52	$FeCO_3$	
K_2O	2.80	K_2CO_3	
Fe_2O_3	2.63	$FeCO_3$	
	21.76	All carbonates	

Table 8: Some minerals carbonate composition and energy values.

5.2.1. Merits and Demerits of Mineral as CO_2 Storage Option

There are three advantages that are generally given for mineral Storage of CO_2.

- Carbonates have a lower energy state than CO_2. Therefore, at least theoretically, the process not only requires no energy inputs, but also can actually produce energy.

A more detailed explanation from Yegulalp *et al.* (2000) follows:

The carbonation reaction can be shown by the simple reaction of binary oxides, MgO and CaO. These reactions are exothermic.

$$CaO + CO_2 \longrightarrow CaCO_3 \quad + 179\ kJ/mole$$

$$MgO + CO_2 \longrightarrow MgCO_3 \quad + 118\ kJ/mole$$

Even compared to the heat released in the combustion of carbon (394 kJ/mole), these reactions release substantial heat. In nature, however, calcium and magnesium are rarely available as binary oxides. They are found typically as calcium and magnesium silicates. The carbonation reaction is still exothermic for common calcium and magnesium bearing minerals. In such cases however, the heat release is reduced. As an example consider the carbonation reactions of forsterite and serpentine. For forsterite and serpentine respectively:

$$\tfrac{1}{2}Mg_2SiO_4 + CO_2 \longrightarrow MgCO_3 + \tfrac{1}{2}SiO_2 \quad + 95kJ/mole$$

$$1/3Mg_3Si_2O_5(OH)_4 + CO_2 \rightarrow MgCO_3 + 2/3SiO_2 + 2/3H_2O + 64kJ/mole$$

Both of these reactions are favoured at low temperatures. In nature magnesite and silica are common in serpentinized ultramafic rocks. Their formation is due to natural CO_2–rich fluids percolating through mineral deposits. Magnesite is stable and not likely to release the bound CO_2 again.

- The raw materials are abundant. Once again from Yegulalp *et al.* (2000):

Calcium and magnesium carbonates are solid which is desirable in above ground disposal. The materials formed can be stored at the serpentine mine as landfill and will not leave the disposal site. Magnesium proved more attractive since there are large deposits of magnesium rich minerals. Peridotite, and serpentinized peridotite rocks can have MgO content between 35% and 40% by weight, whereas abundant calcium silicates rarely have more than 12 % to 15% of CaO by weight. In addition, the magnesium silicates are more reactive and are therefore more suitable for above ground carbonation.

The process implies a large mining effort, but the areal extent of the mine is small compared to the coal mine that produces an equivalent amount of coal. Overburden on serpentinite rock is generally insignificantly small and the minerals occur in thick layers rather than thin seams. Nevertheless, the mass of material required is larger by a factor of six than the mass of coal that is used as fuel. As a result, the formation of carbonate will have to be performed at the mine site and the resulting silica and carbonates will be stored in the mine. Since volumes increase in the process, some modification of the local terrain's profile is unavoidable.

Mining costs appear to be quite low. The mining is similar to copper mining and the amount of peridotite required for a GW power plant is small compared to the amount of ore mined in a large copper mine. Cost estimates, based on other mining operations suggest a cost of about $8 per ton of CO_2.

- This is the only form of carbon sequestration that is "permanent".

The other major proposed storage options (geologic and ocean) have the potential for leakage over time. Leakage can make the sequestration ineffective if the CO_2 returns too quickly to the atmosphere. In addition, leaks may present other environmental, health, or safety issues. Mineral sequestration has none of these concerns.

The only issue that remain a major challenge with mineral storage is that further investigations are required, as in the case with other storage options.

5.3 References and Further Readings

http://www.netl.doe.gov/technologies/carbon_seq/index.html. The site includes a wide variety of information, including links to newsletters, journals, papers, news reports, a video and program plans. For a comprehensive overview of current activities and plans, link to the Carbon Sequestration Technology Roadmap and Program Plan.

http://www.netl.doe.gov/technologies/carbon_seq/refshelf/refshelf.html. See also, the web site of the MIDCARB research consortium (Midcontinent Interactive Digital Carbon Atlas and Relational DataBase) which presents the results of its research evaluating the capacity for geologic sequestration in the member States Illinois, Indiana, Kansas, Kentucky and Ohio at: http://www.MIDCARB.org.

http://www.ieagreen.org.uk/ and companion site focusing on carbon capture and storage at: http://www.co2captureandstorage.info/.

Burying the problem. Canadian Geographic magazine

IEA Greenhouse Gas R&D Programme The Orchard Business Center, Stoke Orchard, Cheltenham, Glos. GL52 7RZ, United Kingdom. www.ieagreen.org.uk

IPCC, 2005. IPCC special report on Carbon dioxide Cpture and Storage prepared by working group III of the Intergovernmental Panel on Climate Change

Herzog, H. J. 2001. What Future for Carbon Capture and Sequestration?. Environmental Science and Technology, 35(7):148A-153A.

Houghton J.T., Meira Filho L.G., Callander B. A., Harris N., Kattenberg A,. and Maskell K. "Climate Change 1995. The Science of Climate Change". Cambridge University Press 1996.

Yegulalp T.M., K.S. Lackner and H.J. Ziock, "A Review of Emerging Technologies for

Sustainable Use of Coal for Power Generation," presented at Sixth International Symposium on Environmental Issues and Waste

Management in Energy and Mineral Production, Calgary, Alberta,

Canada, May 30-June 2 (2000).

CHAPTER SIX: ECONOMIC AND POLICY ISSUES

6.1 Commercial Prospects

The key issues that prompted CCS to be considered as a commercially viable option were partly discussed in some of the previous chapters; therefore, the current chapter serves only as an extension.

One of the major potential benefits of CCS is utilising the gas for enhanced oil recovery (EOR). The method increases the amount of oil that is recovered from an underground oil reservoir. By pumping CO_2 into an oil reservoir, previously unrecoverable oil is pushed up through ideal displacement to where the oil can be reached. The U.S. Department of Energy (DOE) reported that this can produce an additional of 30 to 60 percent of the original amount of recoverable oil once all the recoverable oil has been reached, and the depleted reservoir can act as a permanent storage site for the CO_2 (MGA, 2009). Galadima and Garba (2008) reported that in the Permian basin of Texas United States, CO_2 has been injected into over 11,000 oil wells for tertiary recovery since late 1970's and the technology already accounted for over 15% of annual oil and gas production in the region. A commercial success of the technology is the "Sleipner

Project", where the gas is compressed and pumped into 200m thick Utsira sandstone layer of the Sleipner oil and gas field in the North Sea, Norway (Herzog, 2001). About 1 million metric tonnes of CO_2 have been sequestrated annually since October 1996, and the investment was paid back in about one and- half years only (Herzog, 2001). Similarly, the Weyburn-CO_2 monitoring project, Canada accounts for about $30 million of gross revenues to the gasification plant's cash flow each year (Haefeli et al., 2004). Up to 20 million tonnes of CO_2 would be injected and permanently stored, leading to at least 130 million barrels of additional oil recovery for the project life cycle.

For oil and gas companies involved in methane exploitation from coal beds, studies indicated that up to 173-203 billion tonnes of the gas could be injected for enhanced coal bed methane recovery (ECBMR) worldwide. In developing countries like Nigeria, the current proven oil and gas reserves (36.22bbls & 184tcf) were projected to last for the next 40 years (Galadima etal., 2009), indicating that future CO_2-enhanced oil recovery is necessary.

A potentially useful way of dealing with industrial sources of CO_2 is to convert it into hydrocarbons where it can be stored or reused as fuel or to make plastics. There are a number of projects investigating this possibility. Currently, biofuels represent the other potentially carbon neutral jet fuel available. Carbon dioxide scrubbing variants exist based on potassium carbonates which can be used to create liquid fuels. Although the creation of fuel from atmospheric CO_2 is not a geoengineering technique, nor does it actually function as green house gas remediation, it nevertheless is potentially very useful in the creation of a low carbon economy, as transport fuels, especially aviation fuel, are currently hard to make other than by using fossil fuels. A proven process to produce a hydrocarbon is to make methanol. Methanol is rather easily synthesised from CO_2 and H_2.

$$CO_2 + H_2 \rightarrow \text{Methanol}$$

If CO_2 is heated to 2400oC, it splits into carbon monoxide and oxygen. The Fischer-Tropsch process can then be used to convert the CO into hydrocarbons. The required temperature can be achieved by using a chamber containing a mirror to focus sunlight on the gas.

$$CO_2 \rightarrow CO \rightarrow \text{Hydrocarbons}$$

In addition to all these, the captured CO_2 can be used for many other industrial and house hold applications such as production of polymers and plastics, carbonated drinks and soda water, electric fire extinguishers, urea, industrial welding, as a promising solvent for lipophilic organic compounds, organochlorins, in the removal of caffeine from coffee and as an industrial gas laser (CO_2-laser). These therefore form a chance for many oil and gas companies to diversify their investments by establishing related industries.

6.2 Some Major CCS Projects

There are currently about 90 CCS projects under observations (although some are at the initiation stage) that sequester huge amount of the harmful CO_2 gas, indicating how successful the technology will go as a climate change mitigation option by 2030 and beyond. It is beyond the scope of this book to discuss all the projects, however, a substantial effort had been made to cover the ones that have attracted global attention.

6.2.1. The Weyburn Project, Canada

Although CO_2 has been injected into oil and/or gas reservoirs for enhanced recovery since 1970's, the monitoring of its science and technology had received only partial concern during that period. The Weyburn project (figure 32), Canada was among the first monitoring projects in the global record that kicked up in 2000. The project is operated in conjunction with two billion-dollar commercial CO_2 floods in Saskatchewan, Canada, where huge volumes of the gas are captured from an industrial source and injected to revive oil production. It specifically targeted at scientific, geological and economic feasibility of storing CO_2 in geological fields, especially depleted oil reservoirs (i.e. like the Weyburn field). Emphasis was given to the mechanism of CO_2 up take, distribution and containment and permanent storage within the reservoirs. The outcome was therefore expected to form a basis for selecting other storage sites. The first phase of the 11-year $85 million international project covers the period 2000 to 2004. In 2000, CO_2 supplied by Dakota Gasification Company, USA from the Great Plains synfuels plant was injected at the rate of 5000 t/d, before it was later increased to 6300 t/d in 2002. During this period an additional oil production of over 5000 barrels/day was achieved. The project's Final Phase (2005-2011) is building on the successes of the First Phase (2000 – 2004), to deliver the framework necessary to encourage implementation of CO_2 geological storage on a worldwide scale. Based on preliminary results of the first phase, the natural geological setting appears to be highly suitable for long-term CO_2 geological storage. This outcome formed the most complete, comprehensive, peer-reviewed data set in the world for CO_2 geological storage. However, additional research is needed to further develop and refine CO_2 monitoring and verification technologies. Consequent to the success of the first phase, the final phase continues to utilize scientific experts from every corner of the globe to further develop and build the most scrutinized CO_2 geological storage data set in the world. The outcome is expected to address several important issues. This will include technical components (including site characterisation, monitoring and verification, wellbore integrity and performance assessment), and policy components (including regulatory issues, public communication and outreach, and business environment).

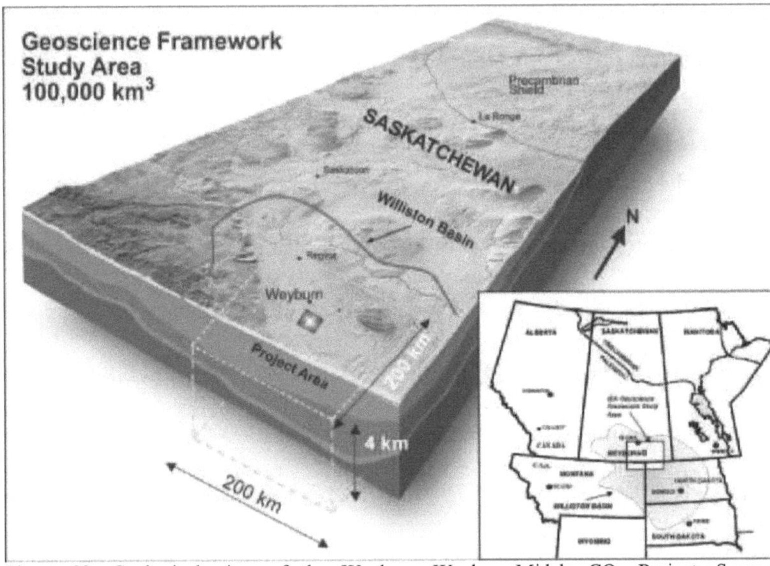

Figure 32: Geological view of the Weyburn Weyburn-Midale CO_2 Project. Source: http://www.zeroco2.no/projects/weyburn-midale-co2-project .

6.2.2. Sleipner CO_2 Injection Project, Norway

The Sleipner project started in 1996, earlier than the Weyburn project, and can therefore be considered the first CO_2 injection and monitoring project. The project is operated by Statoil Hydro at the Utsira Gas Field of the Norwegian sector of the popular North Sea. The Utsira Formation is a deep saline reservoir 8000-1000 meters (2600-3300ft) below the sea floor. The natural gas produced from the Sleipner West field contains up to 9% CO_2, however, in order to meet the required export specifications and the customers' requirements, this has to be reduced to a maximum of 2.5%. The CO_2 is removed from the produced hydrocarbons at an offshore platform before being pumped back into the ground and the hydrocarbons piped to land. Had this process not been adopted, and the CO_2 produced been allowed to escape to the atmosphere, the licensees of the Sleipner West field (figure 33) would have had to pay NOK 1 million/day in Norwegian CO_2 taxes. Statoil has so far stored 8-10MT of CO_2. There is no evidence of CO_2 leakage and the CO_2 remains in situ. CO_2 capture is done using amine technology. Injection currently costs $17 US/Ton CO_2 (http://sequestration.mit.edu/tools/projects/sleipner.html). It has been estimated that the Utsira Formation is capable of storing 600 billion tons of CO_2. 3D seismic monitoring of the CO_2 injection into the field shows that there is no leakage of the CO_2 into other horizons. Like the Weyburn project, the Sleipner project had increasingly contributed to the understanding of CO_2 sequestration, especially to the EU geologists and decision makers.

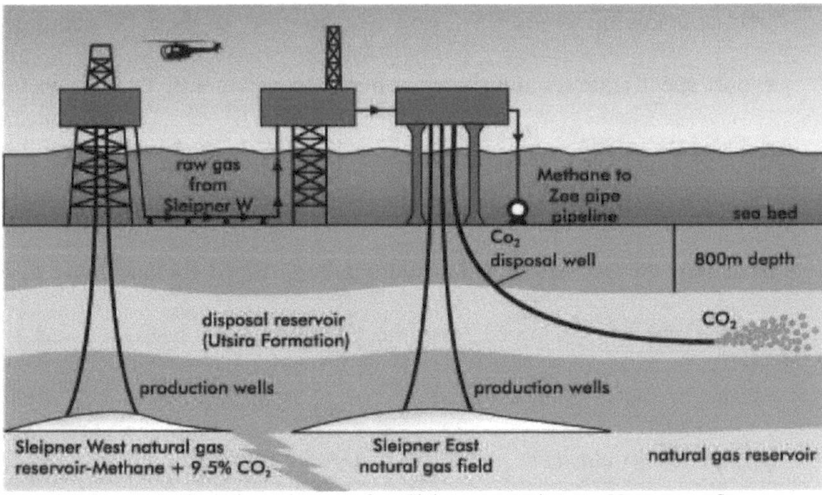

Figure 33: Geological view of Sleipner project, Norway. Source: http://www.co2crc.com.au/publications/all_factsheets.html

6.2.3. In Salah CCS Project, Algeria

The In Salah Carbon Capture and Storage project located in Algeria, Africa (figure 34) and operated jointly by Sonatrach (35%), BP (33%) and Statoil (32%), is the first large scale climate change mitigation project in the continent (but one of the largest in the world), following a response to the UNFCCC call to oil companies on reducing CO_2 emissions. It is an Industrial-scale demonstration of CO_2 geological storage, in which 1 million tons per year of CO_2 will be geologically stored in the onshore field. It was estimated that about 17 million tons of CO_2 will be re-injected during the whole life of the project. CCS reduces GHG emissions of the project by 60%. This is equivalent to take 250 000 cars off road or 200 km^2 of forests. The projected project cost is \$100 million based on CO_2 capture & storage cost of \$6 per tonne. According to the operators the objectives of the project include among others, the followings (Haddadji, 2006).

- First, to demonstrate to stakeholders that industrial-scale geological storage of CO_2 is a viable greenhouse gas (GHG) mitigation option.
- Second, to assure people that secure geological storage of CO_2 can be cost-effectively verified and that long-term assurance can be provided by short-term monitoring; and
- Third, to set precedents for regulating and verifying geological storage of CO_2 - ultimately to allow eligibility for Clean Development Mechanism (CDM).

The project that started in 2004, demonstrated that complexity and project size are not obstacles to achieving the extraordinary outcome, with large scale atmospheric disposal of CO_2 no longer seen as an acceptable option (Haddadji, 2006). In future the project is expected to lead in setting precedents for monitoring, regulation and verification of geological CO_2 storage and establish CCS as eligible for Kyoto Protocol Clean Development Mechanism.

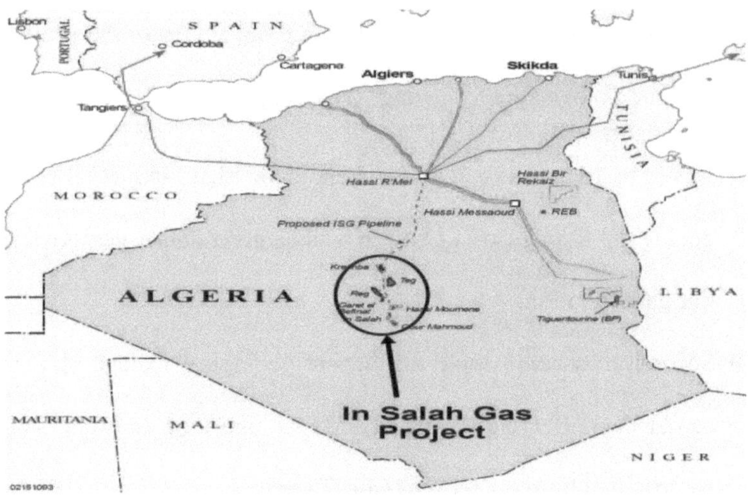

Figure 34: Geographical illustration of the location of In Salah project, Algeria.

6.2.4. Snøhvit CO_2 Project - Barents Sea, Norway

Snøhvit CO_2 Project - Barents Sea, Norway (figure 35) is a pre-combustion capture project proponented by Statoil ASA, Petoro AS (Norwegian state direct interest), Total E&P Norge AS, GDF Suez E&P Norge AS, Norsk Hydro and Hess Norge. Gas production started in October 2007 and currently produces approximately 7 billion cubic metres per annum. Carbon dioxide (CO_2) is removed from the gas stream and piped about 154 km back to the field for injection through a dedicated well. Around 0.7 million tonnes per annum of CO_2 (which equals the emission volume from 280,000 cars) have been safely injected and stored in the Tubåen sandstone (some 2600 metres beneath the seabed and about 45-75 metres thick) since April 2008.

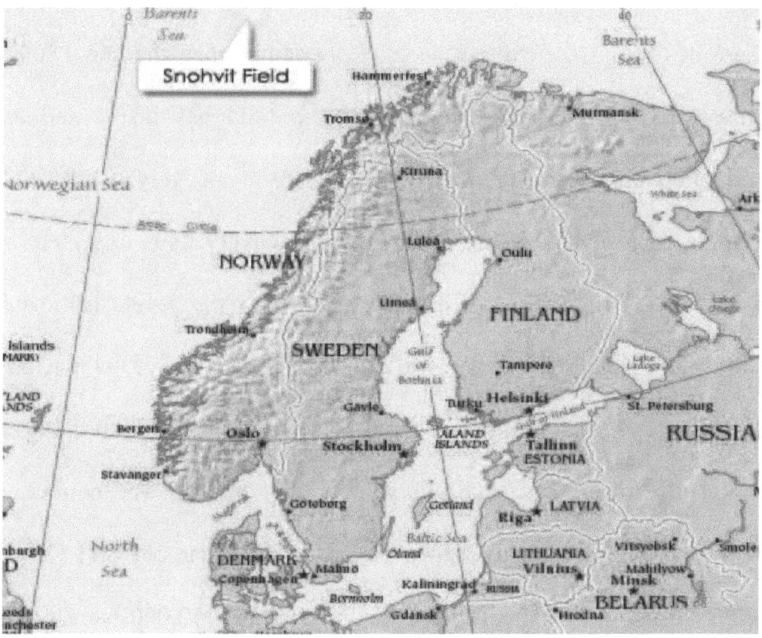

Figure 35: Location of the Snohvit Field, Barent Sea Norway. Courtesy of RIGZONE (2011).

6.2.5.0 Salt Creek Enhanced Oil Recovery, Natrona County, Wyoming, USA.

Operated by Anadarko Petroleum Corporation, the EOR project is fully operational. Anadarko has injected 5.12 billion cubic metres of carbon dioxide into the field as part of a project to tease more oil from the field and in the process sequester a greenhouse gas that would otherwise have to be discharged into the atmosphere (Galadima and Garba, 2011).

The Salt Creek oil field has been in operation since the late 1800s. Since early 1971, water flooding of the field has been used as production improvement method. Enhanced oil recovery using CO2 flooding represents a tertiary form of oil recovery. It is expected to significantly improve production compared to the levels achieved with water flooding alone (HPC, 2009). During the 2004 period, Anadarko Petroleum initiated the first phase of the $200 million carbon dioxide (CO_2) enhanced oil recovery project. The company uses Wyoming-produced CO_2 to flood the 100-year old Salt Creek field and gradually increased production to nearly 30,000 barrels per day by 2008. Production could continue 10 to 20 years to sweep an additional 150 million barrels of oil from the Salt Creek Field. The project is recognised as one of the largest CO_2 oil recovery and geological-sequestration projects of its kind in the world. The project currently sequesters enough CO_2 each day to offset the equivalent emissions of more than 600,000 cars.

6.3 Some Global CCS Policies

Different countries have different approaches toward implementing and utilising CCS. However, the overall objectives could be summed to focus on exploring the technology as commercial tool and climate change mitigation option.

6.3.1 The European Approach

In the European countries, the major coal users such as United Kingdom, Holland (Netherlands), Spain, Poland and Germany have indicated strong commitment to CCS projects in line with the European Economic Recovery Plan (EERP). However, in countries such as Greece, which is a persistent coal user as well (i.e up to 52% of its electricity generation is coal dependant), CCS was strongly opposed on costs ground (Tindale, 2011).

After long time debate, the European Parliament emphasized on inserting a regulatory limit to the amount of greenhouse gases that could be emitted into the Industrial Emissions Directive, 2009. This implies ensuring a strong commitment to CCS approach. However, the effort was only successful in allowing each member country to set its own regulations. Some countries like UK have already made some progress. The European Union Emissions Trading Scheme (ETS) has been in operation since 2005, the largest such capture-and-trade program in the world, covering power and industrial sectors. The price that the ETS has put on a metric ton of carbon emission has varied considerably from €30 at its peak to a low of fractions of €1. A number of literatures consider CCS approach a good commercial opportunity for the Europe in addition to climate change mitigation. However, Environmental activists and local people living in the proposed storage sites opposed the motion on the grounds of safety.

6.3.2 The United States Approach

The United States have a different approach compared to the European Union countries. The Regional Greenhouse Gas Initiative

(RGGI) recently started in 2009 in the Northeast between ten states to limit the emissions of the power sector. The value of RGGI's permits have ranged from a high of $3.51 per (short) ton of CO_2 to a low of $1.87 per ton of CO_2. Major legislation is currently under consideration by the United States Congress to establish a nationwide capture-and-trade program. In Congress, H.R. 2545, the American Clean Energy and Security Act of 2009 (ACES/axman-Markey) has passed the House of Representatives and would establish a capture-and-trade program for CO_2. At the beginning of 2010, S. 1733 the Clean Energy Jobs and American Power Act are being considered in the Senate. The major component of both of these programs is an economy-wide capture-and-trade program, including the electric power sector (Shu, 2010).

A capture-and-trade program would have impacts similar to a carbon tax. For capture-and-trade, a CO_2 emissions target is set and the market determines the carbon price. For a carbon tax, a carbon price is administratively set and the market determines the level of CO_2 emissions. Another difference is that in a capture-and-trade program, policy makers decide how to allocate emissions permits while in a carbon tax policy makers decide how to allocate revenue from the emissions tax (Shu, 2010).

CCS projects are currently ongoing in some key parts of the United States, Salt Creek Enhanced Oil Recovery; Natrona County, Wyoming, USA is one of the common successful examples. No one can explain commercial success of CCS without citing example with EOR project at the Permian Basin Texas, where CO_2 injections remained the key tertiary recovery option. The major challenges of CCS in the USA are similar to those in the European countries.

6.3.3 The China Approach

China is currently the largest emitter of CO_2 from coal combustion, with the sector becoming rapidly industrialised. In line with these, the International Energy Agency (IEA) states that, CCS in China will have to supply 20-25% of total emissions reductions, and over 60% of those reductions will need be applied to coal-fired power plants (Morse et al., 2009). This implies the Chinese government will require $25 to 30 billion annually for CCS projects. Although some projects are already in place, some authors accused China of showing only partial commitment at both national and international scales. Morse et al. (2009) have identified the following limitations.

- Technological uncertainty, especially regarding performance of capture technologies at scale.
- High capture, transportation and storage costs.
- Regulatory uncertainty including over liability for sequestered CO_2.
- Lack of clear carbon policy that could provide a revenue stream for CO_2 capture.

6.4 References and Further Readings

Haddadji, R (2006) EU-OPEC Roundtable on Carbon Dioxide Capture & Storage. Session 2: Prospects of CCS in OPEC Member Countries and the EU. The In-Salah CCS experience Sonatrach, Algeria. The First International Conference on the Clean Development Mechanism. Riyadh, Saudi Arabia (The Intercontinental Hotel) 26-28 Shaaban, 1427, 19-21 September 2006.

http://www.opec.org/opec_web/static_files_project/media/downloads/press_room/HaddadjiSonatrach_Algeria.pdf

Farley, M (2009) CCS Technology and capture-ready plants. Energy World, A publication of

Energy Institute (EI), January, 2009. Pages 2-3.

Galadima, A., Kabiru, A., Garba, Z.N., Abubakar, A (2009) Carbon Capture and Storage (CCS): A future challenge to Nigeria. A paper presented during the 7th CHEMCLASS Conference, 11th June ,2009, A,B.U. Zaria –Nigeria.

Galadima, A. & Garba, Z.N. (2008) Carbon Capture and Storage in Niugeria: Fundamental Science and Potential Implementation Risks. Science World Journal, 3(2)95-99.

Galadima, A. and Garba, Z.N (2011) Recent Issues in Environmental Science: "Including incidences and reports from Nigeria"-A book, Lap Lambert Academic Publishing, Germany. ISBN: 978-3-8454-2915-1.

Haefeli, S., Bosi, M., Philibert, C (2004) Carbon dioxide capture and storage issues- Accounting and Baselines under United Nations Framework Convention on Climate Change (UNFCCC). IEA information paper, Paris, May, 2004.

Herzog, H.J (2001) What future for carbon capture and sequestration?. Environmental Science and Technology, 35(7) 148A-153A.

Herzog, H. And Golomb, D(2005) Crabon capture and storage from fossil fuels use. Massachusetts Institute of Technology, Cambridge, Massachusetts, United States. Article

Number: NRGY: 00422.

Howell Petroleum Corporation (HPC, 2009) Greenhouse Gas Emission Reduction Creation Report for the Salt Creek Enhanced Oil Recovery (EOR) Project for CY 2008. Howell Petroleum Corporation Natrona County, Wyoming, February 2009.

Howell Petroleum Corporation (HPC, 2004) Greenhouse Gas Emission Reduction Protocol for the Salt Creek Enhanced Oil Recovery (EOR) Project. Howell Petroleum Corporation Natrona County, Wyoming, August 2004, Revision 4.

In Salah Gas Project (2011) The overview: http://www.insalahco2.com/index.php/en/co2-storage-at-in-salah/monitoring-technologies.html#overview

MGA (2009) Carbon capture and storage and Enhanced oil recovery, fact sheet. Available

From http://www.midwesterngovernors.org , Last visited in March, 2009.

Morse, R.K., Rai, V and He , G (2009) The Real Drivers of Carbon Capture and Storage in China and implications for climate policy. PROGRAM ON ENERGY AND SUSTAINABLE DEVELOPMENT. Working Paper #88, August, 2009.

RIGZONE(2011) Map: Snohvit Field.
http://www.rigzone.com/news/image_detail.asp?img_id=2835

Shu, G (2010) Economics and Policies for Carbon Capture and Sequestration in the Western United States: A Marginal Cost Analysis of Potential Power Plant Deployment. MSc Thesis, Department of Urban Studies and Planning, Massachusetts Institute of Technology, 2010.

Tindale, S (2011) Smart and Sustainable Growth. Carbon Capture and Storage: EU advancing, but not fast enough. TGAE, 2011.

Kolibri 360 GmbH
Ferdinand-Jühlke-Straße 7
99095 Erfurt, Deutschland
produktsicherheit@kolibri360.de

Druck:
CPI Druckdienstleistungen GmbH
im Auftrag der
Kolibri 360 GmbH
Ferdinand-Jühlke-Str. 7
99095 Erfurt